The evolution of IBP

THE INTERNATIONAL BIOLOGICAL PROGRAMME

The International Biological Programme was established by the International Council of Scientific Unions in 1964 as a counterpart of the International Geophysical Year. The subject of the IBP was defined as 'The Biological Basis of Productivity and Human Welfare', and the reason for its establishment was recognition that the rapidly increasing human population called for a better understanding of the environment as a basis for the rational management of natural resources. This could be achieved only on the basis of scientific knowledge, which in many fields of biology and in many parts of the world was felt to be inadequate. At the same time it was recognised that human activities were creating rapid and comprehensive changes in the environment. Thus, in terms of human welfare, the reason for the IBP lay in its promotion of basic knowledge relevant to the needs of man.

The IBP provided the first occasion on which biologists throughout the world were challenged to work together for a common cause. It involved an integrated and concerted examination of a wide range of problems. The Programme was co-ordinated through a series of seven sections representing the major subject areas of research. Four of these sections were concerned with the study of biological productivity on land, in freshwater, and in the seas, together with the processes of photosynthesis and nitrogen fixation. Three sections were concerned with adaptability of human populations, conservation of ecosystems and the use of biological resources.

After a decade of work, the Programme terminated in June 1974 and this series of volumes brings together, in the form of syntheses, the results of national and international activities.

IBP personalities in 1974 *from left to right*

Extreme back: F. Bourlière (France) *President*

Back row: W. F. Blair (USA) *Vice-President*; M. J. Dunbar (Canada) *PM Convener*; J. S. Weiner (UK) *HA Convener*; H. Tamiya (Japan) *Vice-President*; R. W. J. Keay (UK) *Chairman of Finance Committee*

Front row: E. M. Nicholson (UK) *CT Convener*; G. K. Davis (USA) *UM Convener*; C. H. Waddington (UK) *former Vice-President*; J. G. Baer (Switzerland) *President Emeritus*; G. Montalenti (Italy) *former Vice-President, Chairman: Planning Committee*; Livia Tonolli (Italy) *PF Convener*; E. B. Worthington (UK) *Scientific Director*; Sir Otto Frankel (Australia) *Vice-President*

Inset left: J. B. Cragg (Canada) *PT Convener*

Inset right: I. Málek (Czechoslovakia) *Vice-President*

INTERNATIONAL BIOLOGICAL PROGRAMME 1

The evolution of IBP

EDITED BY

E. B. Worthington

Scientific Director
International Biological Programme

CAMBRIDGE UNIVERSITY PRESS

CAMBRIDGE

LONDON · NEW YORK · MELBOURNE

CAMBRIDGE UNIVERSITY PRESS
Cambridge, New York, Melbourne, Madrid, Cape Town, Singapore, São Paulo, Delhi

Cambridge University Press
The Edinburgh Building, Cambridge CB2 8RU, UK

Published in the United States of America by Cambridge University Press, New York

www.cambridge.org
Information on this title: www.cambridge.org/9780521116114

© Cambridge University Press 1975

First published 1975
This digitally printed version 2009

A catalogue record for this publication is available from the British Library

Library of Congress Catalogue Card Number: 75–2722

ISBN 978-0-521-20736-2 hardback
ISBN 978-0-521-11611-4 paperback

Contents

Contents

Table des matières

Table des matières

Содержание

Содержание

Contenido

Contenido

Illustrations

Foreword

The purpose of this book is to set the scene for the thirty-five or so others which will follow in the IBP synthesis series. The International Biological Programme as a whole is explained – how it was conceived, how initiated, how conducted, and how completed, the whole programme taking the form of an evolving process which occupied some five years of preliminary discussion and planning commencing in 1959, followed by a decade of work, 1964–74. Some background data are presented in the Appendices, which may be useful for reference with subsequent more specialised volumes.

The publication of this series is an ambitious exercise on the part of the many editors and several hundred contributors, as well as on the part of the Cambridge University Press. To produce so much information quickly, in the course of two or three years, might be thought to be over-hasty, especially when some of the research on which it is based still continues. However, so much work concerning the environment is now being initiated, both in the governmental and non-governmental spheres, that we thought it important to produce the syntheses of IBP results now, when they can be brought to bear immediately on the new plans and activities, rather than to wait several years until all results could be integrated comprehensively. As a result of this rather quick action, some of the volumes may suffer, but we believe that the world as a whole will benefit.

The current 'Environmental Revolution', which is still gaining momentum, has occurred during the same period as IBP. Several thousand biologists have been associated with IBP and they have included scientific leaders in many countries. Not only have they contributed to knowledge, but their writings and sayings on such matters as the ecology of natural resources, on population problems and on conservation, have reached out from the former narrow circle of environmental scientists to economists, historians, politicians and to the general public. This process has been stimulated by the expanding circle of contacts between human and environmental scientists which forms what is sometimes referred to as the 'IBP network', the creation of which some would say is the most important achievement of this programme. IBP would not claim responsibility for the environmental revolution, which has stimulated such world activities as the Biosphere Conference at Paris in 1968 which led to the Man and Biosphere Programme, and the UN Conference on the Human Environment at Stockholm in 1972 which led to the creation of UNEP. Nevertheless, there is no doubt that the kind of thinking among biologists which resulted in the creation and prosecution of IBP, also contributed to these great developments. The lead is now being taken by governments acting individually, regionally and on the world scale, and these changes in outlook have, of course, influenced IBP itself, whose own thinking and planning has changed significantly during the decade.

From the experience of IBP, which is in some respects unique in scientific co-operation, a number of lessons can be learned which may be of value to future comparable exercises. These lessons will become apparent in the chapters which follow, but it may be useful to draw attention to a few of them here.

One relates to the funding of national research work by participating countries. Although IBP, like all other ICSU enterprises, was based primarily on the non-governmental sector of science, it inevitably involved a substantial number of government biologists. The funding of projects within the programme, depended ultimately in large measure on government finance, funds being provided usually through the agency of national academies, universities or research councils. In the developing countries of the world, which do not as yet have very much research capacity outside government institutes, it was hoped that governments would support the programme also, but in spite of good planning among local scientists, few of the developing countries were able to launch and finance substantial national programmes on their own account, primarily owing to the difficulty of funding, and this was a disappointment. In new programmes of intergovernmental character this problem of finance may be less acute, but at the same time it may not be so easy as in IBP to sustain the enthusiasm and voluntary effort from non-government scientists which is such a feature of ICSU activities.

With hindsight, one may doubt the wisdom of plunging into a programme so large as IBP with so little guaranteed finance, in particular with no central research fund, from which grant aid could have been issued from time to time. Some of the initial planning indeed tended to assume that such a central fund would be created and in consequence disappointment was experienced.

Another lesson is that, with voluntary participation providing the programme's strength, the objectives need to be broadly rather than narrowly defined in order to attract a sufficiency of scientists. Moreover, the organisation for co-ordinating the whole must be light rather than heavy. Nevertheless, it is essential to have a central executive capable of carrying out the decisions of the controlling committee, and the various sections into which any programme is divided must likewise be properly staffed. Such staff, whether on an honorary or paid basis, would have difficulty in being effective unless they included respected scientists in their own right, as well as competent administrators.

Concerning methods of research, in some well established disciplines, particularly in the physical and chemical sciences, it may be both feasible and desirable to standardise methods so that the results are directly comparable wherever and by whomsoever they are obtained; but this is not the case in many branches of environmental biology. The subjects are still young and are evolving rapidly through the improvement of their methods, so that to standardise would be to stultify. Nevertheless, there is always great need, which IBP has endeavoured to fulfil, for guides in methodology, particularly for those workers in isolated conditions who want some guarantee that their results will be comparable with others.

In the overall administration and structuring of the programme the interdisciplinary approach to all projects was prominent, but it may be that there was too much emphasis at an early stage on a solid structure based on seven sections. Probably, at the level of ecological thinking which prevailed in the early 1960s, this was the only way to get the programme into action, because the various disciplines which we wanted to combine were distinct. There was little common language, for example, between the human sciences and the environmental sciences, and even aquatic biologists and terrestrial biologists used different vocabularies. Whether at that time a better start could have been made is debatable, but the rather rigid sectional structure mitigated against truly interdisciplinary results.

Once IBP got under way a remarkable spirit of enthusiasm pervaded nearly all participants. Nevertheless, the programme has not had an entirely smooth passage. Some problems were added to by the untimely death of several key biologists who are named in the appropriate chapters. There were difficulties occasioned by local politics, as well as of funding. The Central Office in London was broken into by thieves no less than five times and on one occasion, being, it seems, disappointed at finding so little in the cash box, they set fire to the storeroom before departure. The contents of several rooms smouldered in a deoxygenated atmosphere for some hours before discovery, when water added to the confusion.

Foreword

The office was at work again next morning but was not back to normal for months, and some irreplaceable records as well as a large quantity of publications were destroyed.

Finally I would express deep appreciation, on behalf of the Officers and Members of the former Special Committee for the IBP, and the current Publications Committee, both of which were set up by ICSU, to all those who have helped in or participated in the programme, nationally or internationally. These include not only active scientists but officers of United Nations agencies, of ICSU and its Unions, of governments, national academies and foundations. In particular I would thank those leaders of IBP who have contributed to this volume and are named in the various chapters; the Syndics and staff of the Cambridge University Press who are most helpful in the preparation and publication of this series of volumes; the past and present members of SCIBP's Central and Sectional Offices, Gina Douglas and Sue Darell-Brown who compiled most of the appendices and Diana Atkinson who suffered the burden of secretarial work in preparing this book.

To sum up it may be claimed that the main achievement of IBP has been to get away from the static approach of description, as applied to the human species and to his environment, and to work towards a dynamic and functional approach. It is fair comment to state that during the past decade the face of ecology as a science has been changed.

E. Barton Worthington

Sussex, July 1974

Postscript

When this volume was in proof the sad and untimely death occurred on 21 February 1975 of Prof. Jean G. Baer of Neuchatel, Switzerland. Among the many distinguished episodes of his career as an International Zoologist, his work as one of the 'Founding Fathers' and later First President of IBP was of high importance.

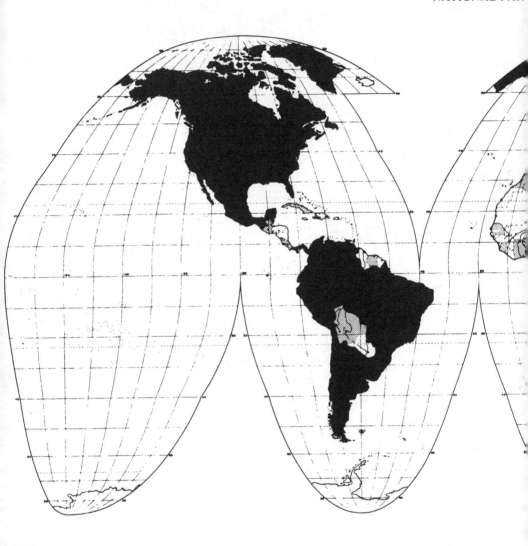

Countries with IBP National Committees

Countries in touch with IBP through correspondents or contributing projects

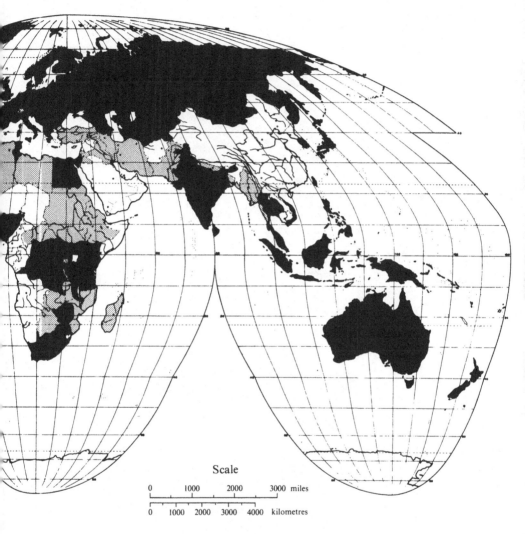

Scale

0	1000	2000	3000 miles

0	1000	2000	3000	4000	kilometres

Goode Homolosine Equal-area Projection

Prepared by Henry M. Leppard
© 1961 by The University of Chicago

1. The origin and early history

The two International Polar Years of 1882–3 and 1932–3, followed by the International Geophysical Year (IGY) of 1957–8, had drawn attention to the advantages of international collaboration in geophysical science and how this could often be turned to practical account of advantage to mankind. It is understandable therefore that biologists were stimulated to think of some similar international effort. In the discussions and events just after IGY, during the years 1959–64, which culminated in the actual inauguration of the IBP, three biologists took a leading part, namely Sir Rudolph Peters, who was President of the International Council for Scientific Unions (ICSU) from 1958–61, G. Montalenti and C. H. Waddington, who were successively Presidents of the International Union of Biological Sciences (IUBS). Sir Rudolph has kindly written a brief statement and Waddington has written a personal account of events as he saw them.

Montalenti has declined to write separately, being satisfied that the subject is well covered by the other contributions. For additional information, however, attention is drawn to his papers (1961 and 1966) in which he looks at the origin and early development of the programme, as well as to the account of an early discussion which is summarised in IUBS (1961). In a lighter vein, many of the elements of IBP are well expressed in Montalenti's Christmas card for 1973, as follows:

CANTICO DELLA CREATURE

Laudato sie, mi Signore, cum tucte le tue creature,
spetialmente messer lo frate sole,
lo quale jorna, et allumini per lui;
et èllu è bellu e radiante cum grande splendore;
de Te, Altissimo, porta significatione.

Laudato si', mi Signore, per sora luna e le stelle;
in celo l'ài formate clarite et pretiose et belle.

Laudato si', mi Signore, per frate vento,
et per aere et nubilo et sereno et onne tempo,
per le quale a le tue creature dai sustentamento.

Laudato si', mi Signore, per sor'acqua,
la quale è molto utile, et humele, et pretiosa et casta.

1

Laudato si', mi Signore, per frate focu,
per lo quale ennallumini la nocte,
et eliu è beilu, et jucundo, et robustoso et forte.

Laudato si', mi Signore, per sora nostra matre terra
la quale ne sustenta e governa,
e produce diversi fructi, con coloriti fiori et herba.

Laudato si', mi Signore, per sora nostra morte corporale,
da la quale nullu homo vivente po skappare.
Guai a quilli ke morrano ne le peccata mortali.
Beati quilli che se trovarà ne le tue sanctissime voluntati
ka la morte secunda nol farrà male.

Be praised, my Lord, with all your creatures, especially
master brother sun, who brings day, and you give us
light by him. And he is fair and radiant with a great
shining – he draws his meaning, most high, from you.

Be praised, my Lord, for sister moon and the stars,
in heaven you have made them clear and precious and lovely.

Be praised, my Lord, for brother wind and for the air,
cloudy and fair and in all weathers – by which you give
sustenance to your creatures.

Be praised, my Lord for sister water, who is very
useful and humble and rare and chaste.

Be praised, my Lord, for brother fire, by whom you
illuminate the night, and he is comely and joyful and
vigorous and strong.

Be praised, my Lord, for sister our mother earth, who
maintains and governs us and puts forth different fruits
with coloured flowers and grass.

Be praised, my Lord, for sister our bodily death, from
which no living man can escape; woe to those who die in
mortal sin; blessed are those whom it will find living
by your most holy wishes, for the second death will do
them no harm.

Francesco d'Assisi (1182–1226)

(Translated by G. Kay)

Statement
by Sir Rudolph Peters

20 November 1973

My first activity as President of the International Council of Scientific Unions, was to act as Chairman of an ICSU Bureau meeting which was held in Gonville and Caius College, Cambridge, in March 1959, when F. J. M. Stratton, who had in effect organised ICSU from his college rooms for a number of years previously, was host. An excellent dinner was given to us by the Master of Caius, Neville Mott.

After this meeting, in the train to London, I fell to discussing with the Past President, Lloyd Berkner, and Guiseppe Montalenti, the possibility of having a project in biology similar to the IGY. At first I thought that this might be in the nucleic acid field; but on consulting A. Todd in Cambridge and L. Beadle, then in Oxford, it seemed to be clear that the nucleic acid field was then fully stretched. Nevertheless, I went on with the general idea and on my way to Italy in the spring of 1959 I discussed it with R. Fraser, Secretary of ICSU at our Office in the Hague. I then went on to Naples to see Montalenti, then President of the International Union of Biological Sciences. He was enthusiastic about the idea, and thought that it was particularly important to make observations upon human populations which were more or less still isolated on islands, in valleys, etc., before civilisation stirred these up beyond genetic recognition. Montalenti always supported this and it became part of the final programme.

The idea of a biological programme spread and was discussed at the Lisbon meeting of the Executive Board, at which I was Chairman and at which Montalenti gave a lecture. The rest is well described by Waddington. The narrow field did not appeal to those from the USSR who rightly wished it to be extended to human welfare. At the same time they were extremely critical for at least a year, though they never actually banged the door shut. The position was so bad at one period that I made up my mind that we should proceed with the project with or without the USSR. Fortunately they decided eventually to take part and it became an important part. I think that IBP would have failed without the help of Waddington. It did accomplish what I had hoped – a synthesis of the isolated research efforts of biologists in many fields and even the setting up of new stations for work.

The Evolution of IBP

The Origin
C. H. Waddington

I first came into any close contact with the IBP at the General Assembly of the IUBS held in Amsterdam in July 1961. When I was sent by the Royal Society as one of the British delegates to this Assembly, I knew very little about the IUBS or its work. It was a great surprise to me when towards the end of the meeting I was approached by Paul Weiss on behalf of the Nomination Committee and asked to allow my name to be put forward as the next President of the Union. I was rather reluctant to accept this, but eventually did so. One of the reasons why I was willing to contemplate taking on the job was that my interest had been aroused in the potentialities implied in the proposal to start an International Biological Programme.

The first suggestion that an International Programme on Biology should be launched seems to have originated with Sir Rudolph Peters in the late fifties, when he was President of ICSU. He communicated the idea to G. Montalenti, the then President of IUBS. Montalenti drew up the first at all formal scheme for an IBP, which he presented to the Executive Committee of ICSU at its meeting in Lisbon in 1960. ICSU appointed a preparatory committee, with Montalenti as Convener and containing Engelhardt (USSR), Florkin (Belgium), President of the International Union of Biochemistry, MacIntosh or Lindor Brown (UK), representing the International Union of Physiological Sciences, L. C. Dunn or G. L. Stebbins (USA), the latter being Secretary-General of IUBS, and C. Pantin (UK), zoologist. This preparatory committee, with some co-opted people, held its first meeting in Cambridge, England, in March 1961, and drew up a document about possible subjects to be dealt with by the IBP. They agreed that they would put forward a small number of rather definite projects in fairly restricted areas. Three such areas were proposed: (1) human heredity (Montalenti); (2) plant genetics and breeding (Stebbins); and (3) studies of natural biological communities which are liable to undergo modification or destruction (Baer). The preparatory committee held another meeting in Paris in May 1961; following this, memoranda on the three topics mentioned above were presented to the IUBS Assembly in Amsterdam later that year.

Engelhardt had not been able to attend either the Cambridge meeting in March or the Paris meeting in May of the preparatory committee, but he had been present as a vice-president of ICSU at the Executive Committee meeting of that body in Lisbon. Another Soviet biologist, Kursanov, was a member of the Executive Committee of IUBS and had

4

been present at its meeting in 1960 at Neuchatel. Academicians Engelhardt and Kursanov submitted to the Amsterdam IUBS Assembly a document in which they pointed out that the ICSU Executive at Lisbon had agreed that the title of the IBP should be 'The biological basis of man's welfare'. They argued strongly that the basic social importance of fundamental biology is in its relation to food production and human health, and that therefore the programme 'should be based on investigations leading to the detection, registration and the most effective and rational utilisation of both well known already used, and new, biological resources of plant and animal origin, with the aim to raise the standard of life of mankind'.

At Amsterdam, a morning was devoted to a symposium by four invited speakers on possible topics for the IBP. I was one of those invited to contribute. From what I had learned in the past, I had formed the impression that up to that time the whole enterprise was extremely dubious. There was talk of organising something on a large worldwide scale, comparable to the IGY, but the people putting forward the ideas seemed to have no firm grasp of how it should be financed or organised. Moreover, each sponsor seemed to think that the programme should essentially be devoted to his own speciality – Montalenti to human genetics, Stebbins to plant genetic resources and Baer to nature conservation. My private opinion was that it would probably be most satisfactory to kill the whole thing before it went any further, if this could still be done. However, it seemed rather likely that matters had already gone too far for this to be practical. If, on the other hand, it were to proceed, I felt that the only possible line would be to formulate a programme around something which was indubitably of major social and economic importance for mankind as a whole.

It seemed pretty obvious that the kind of biology in which I was professionally engaged – genetics, epigenetics and so on – did not need, and was not suitable for, any internationally organised co-operative attack of the kind contemplated in an IBP. I therefore felt I had no personal interests at stake, and could attempt to form a judgement from a relatively neutral point of view as to what was the most important contribution biology could make to man. The most attractive field, I thought, was something to do with the way in which solar energy is processed by the biological world into the formation of complex molecules which man can use, as food or otherwise. I gave my contribution to the Amsterdam symposium along these lines. Kursanov took a very similar line, but he, being a professional student of photosynthesis,

5

laid most of his emphasis on studies of different photosynthetic mechanisms. However, he tied these up with the study of general biological productivity. The other speakers certainly did not lay their major emphasis on productivity, but on various other somewhat narrower aspects – for instance, conservation or human heredity.

The Amsterdam Assembly finally passed the resolution that an IBP should be set up and that its main aim should be 'toward the betterment of mankind'. It mentioned three specific areas for action: conservation, human genetics, and improvements in the use of natural resources. A further meeting of a planning committee was arranged to take place early in 1962.

Finding myself as President of IUBS, which was the main union concerned with IBP, I now felt that I had to take a really serious look at what IBP was supposed to be about. In January 1962 I had ready, and circulated to the rest of the Planning Committee, a document under the title 'Notes on the selection of topics for an International Biological Programme'. I suggested that there were three main topics that might be seriously considered as a possible central theme: (1) *Human genetics*. The trouble with this is that although it is easy to deal with characters of relatively minor importance, such as variants and blood groups, the really important subjects (e.g. racial differences, differences in intellectual and other abilities) are both politically highly controversial, and also, at present, lack an adequate scientific understanding. (2) *Human population growth*. I argued that this was unsuitable, both because the facts are already fairly well known and are under intense study by many bodies with far greater financial resources than IUBS (e.g. the Ford Foundation), and again because the implications are the subject of political controversy. This left (3) *Man and ecology*, that is to say the way in which man can modify the natural environment so that it produces with maximum efficiency on a long-term basis the kind of products he can use.

This document formed the starting point for the discussions of the next meeting of the Planning Committee, which was held in Morges, Switzerland, in May 1962. Our host at this meeting was Jean Baer, President of the International Union for the Conservation of Nature (IUCN) as well as Chairman of the Division of Zoology of IUBS. We were meeting, therefore, in a rather 'conservation-orientated' atmosphere. Moreover, there were several other IUCN members present at the meeting (H. Ellenberg, Max Nicholson, and Ed Graham). I found myself in the position of being almost the sole sponsor of the idea that ecology should be looked at as a matter of energy through-put and processing. Kursanov, who would have been entirely on my side, was not able to attend and was replaced by Steinberg of Leningrad, a very nice general zoologist who, however, was not very clued up about production ecology. The main man who really knew about this, in one particular field was Sidney Holt of FAO, a fisheries expert, but his main point was that fishery biology was already very largely internationalised and did not really need much more international organisation – and at any rate, if it did, this should fall to the province of FAO and not of IUBS.

The difficulties of that meeting were increased by the fact that Ledyard Stebbins, the General Secretary of IUBS ,had circulated a questionnaire, based partly on my memorandum and partly on his own ideas, to a few hundred American biologists. Now, general ecology was certainly not one of the strong points of American biology at this time. Such terrestrial ecologists as there were, were either 'wildlife and forestry' oriented, or had particular research projects of their own which took up their full energies (e.g. the Odums). The replies to Stebbins's questionnaire were, therefore, somewhat discouraging, and to my mind almost totally irrelevant. It was clear to me that if we were to organise an IBP concerned with the energetics of different ecosystems we had got to start almost from scratch in insisting that this was indeed the subject we felt ought to be studied. I had come across much of the same attitude in Britain at meetings of the Royal Society committee dealing with the IBP. The general idea amongst biologists at large seemed to be that ecology dealt with a blow-by-blow account of a day in the life of a cockroach, woodlouse or sparrow; and the notion that it could study such questions as what does the ecosystem do with the incident solar energy, tended to be greeted with blank stares of incomprehension – whatever was one talking about?

The Committee finally came up with a structure for the organisation of IBP which is essentially what has been preserved throughout the programme, namely, a set of seven sections or sub-committees: three on biological productivity in terrestrial communities (of these three, one on general productivity, a second on the metabolic processes – mainly biosynthesis and nitrogen fixation – on which this productivity depends, and a third on the conservation of threatened communities); then a fourth on productivity in fresh water; a fifth on productivity in marine communities; a sixth on human adaptability (physiological and genetic); and a seventh on public relations and training. Stebbins was made Convener of the seventh, but in the event each of the more specific

7

sections had to look after its own public relations and training programmes. The seventh section therefore became devoted to more practical problems and was referred to as 'Use and Management'; it aimed to deal with those aspects of applied biology which in one way or another fall just outside the fields of FAO and WHO.

The whole scheme had, of course, still to be sold to the biological world. It was right up the same street as the Russians had suggested, and I suppose took little selling to them, but they hold themselves so much aloof from the rest of the world that it was quite a long time before one knew whether they were playing or not. France and the French-speaking world in general was pretty forthcoming, particularly through the co-operation of F. Bourlière in the field of terrestrial ecology. Sweden with W. Rodhe in freshwater biology, soon followed by Italy with L. Tonolli, also quickly took its place. On the marine side, R. Glover of Edinburgh and several Scandinavians soon picked up the ball.

The really difficult biological communities were those in Britain and particularly those in America, which had very strong traditions of the dominance of physiology, biochemistry and molecular biology. In Britain the dominant medico-physiological establishment, who hold more or less *ex officio* the main official positions in the Royal Society, were sceptical and indifferent, though not definitely antagonistic. The intellectual leaders in molecular biology and genetics were just simply uninterested, but again not really effectively hostile. The only strong British group in the ecological–natural history field was centred round the Nature Conservancy, which had its own sub-committee (the third) in the programme and which was even prepared to think they might in future take productivity more seriously than they had done in the past. Meeting the Royal Society IBP Committee in its early days, I often felt that I was being called upon to invent a new discipline of production ecology off the cuff; but the Royal Society officials – the Executive Secretary, David Martin, and his assistant, Ronald Keay, whose professional fields were science policy – immediately saw the socio-economic importance of IBP and gave it every assistance.

The toughest biological community into which to launch the scheme was that of the United States. The American biological world was not dominated – establishmentwise – by medical physiologists and biochemists to the same extent as the British; but the analytical school of molecular biologists and microbiological geneticists had a far higher status than in Britain and much less hesitation in asserting, in the hearing of government or the academy, that any organism bigger than *E. coli*

8

serves only to confuse the issue. Moreoever, there were schools or personalities in field biology which ranked much higher in the pecking order than any comparable groups in Britain, for instance Dobzhansky and Mayr in evolutionary biology in the field, and even one or two ecologists. Finally, practical ecology in the United States was famous for two 'achievements' – if that is the right word – at the opposite ends of the spectrum: the view that there is enough land to justify simply mining it – leading to the dust bowls of the thirties; and by contrast the development of an extremely elaborate wildlife and National Parks system, which (fairly successfully) aims to produce unspoilt wilderness, rivers full of salmon or forests full of deer within reasonable vacationing distance of the main centres of population. The idea of studying the energy balance of ecosystems had therefore got to thread its way between the adherents of the 'central dogma', who couldn't care less but were apprehensive it might take away some of their public funds, and an opposite party whose line was that 'we *are* field biology, and productivity is not an American problem'.

I had a particularly 'close huddle' with the Americans in May 1963. I had taken myself across to Washington, under no particular official auspices, but with some financial help from Edinburgh University – Sir Edward Appleton, the Principal, didn't quite like giving it to me, but eventually did so. My basic purpose was to interview people in the US science policy establishment in Washington, to find out how their system worked in order to use this knowledge in connection with the Trend Committee's restructuring of British civil science – but this is another story. I also took the chance to see some of the important US biologists concerned with IBP. T. C. Byerly, of the US Department of Agriculture, and head of the Division of Biology in the National Academy, fixed up an interview at which he and about a dozen others were present. Byerly was already fairly favourable, but he was playing it very cool indeed; and many of the others started pretty sceptical about what these Europeans were up to. I had both to convince them that production ecology could be good science, and to pull out all the policy stops, about the starving world, etc. However, Byerly told me some time afterwards that this was one of the key meetings which began to bring the Americans round to the idea that they should play a part. In the early stages, the most helpful American was Edward Graham, who had attended the Rome meeting as a representative of IUCN. But he was a rather mild person, and also, unfortunately, he died before the programme had got very far. He was succeeded by Stanley Cain, also from the

9

wildlife and conservation field. Later the Americans put in charge someone with the reputation of a real thruster, Roger Revelle.

I found the initial hostility in the senior American biologists sufficiently disturbing to write a personal letter to a number of my more influential friends such as Muller, Sonneborn, Sewell Wright, Mayr, Dobzhansky, and Ebert. Ebert's reply (3 June 1963) was typical: '. . . at the AIBS meeting last August, most . . . of the proposals were for warmed-over ecology of the thirties and forties on a worldwide scale . . . I think my statement (made to a group made up to a large extent of conventional systematists and ecologists) had excellent shock value. I believe strongly in a well planned international program . . . which is truly international in character, and which attempts to use the international aspect of the endeavour as a powerful tool, rather than just a gimmick to keep up with the physicists. You can count on my assistance'.

It was not till about 1966 that the real importance of IBP got across to the Americans. In a Plenary Session of the National Research Council, 13 March 1967 (see *Proc. Nat. Acad. Sci.*, 1968, **60**, 1–50), a leading American ecologist, Frederick E. Smith, gave the introductory speech in which he spoke of IBP as 'lifting a minor subject to a position of major status'. Again, 'The participation of many ecologists in IBP . . . will have a profound effect on the orientation of the profession . . . This change of orientation has already begun . . . Within the last year, the sudden support of IBP by a large number of ecologists, after strong initial resistance, has been surprising. The ferment that took place at Williamstown last October had all the excitement of a "happening".' As things subsequently turned out, these manoeuvres paid off handsomely, for the ecological contribution to IBP which the United States made later on exceeded that of any other country and did much to change the face of environmental science.

The other thing I set about doing, early in 1963, was to find suitable staff, and in particular a suitable senior executive. I clearly couldn't and didn't want to run it myself – I had already spent more time on it than I would have liked. Montalenti also could be no more than a President – he hadn't the spare time, or the knowledge or real understanding of the aims of the productivity approach, to guide it when it began to get down to detail. I was not much impressed with the disinterestedness of most of the ecologists who had been associated with the early stages, nearly all of whom were so closely connected with particular lines of work that they had axes to grind. What we needed, I thought, was someone who had considered experience in ecology, and in administration, and who was not likely to try to use the job to build up his own *scientific* reputation – though success would obviously do much for his administrative standing. The best person I could think of was Barton Worthington, who combined long ecological and administrative service in Africa with now being Deputy Director of the Nature Conservancy. I therefore started (February 1963) to suggest to Montalenti that we needed an Executive Secretary, or something of the sort, and that Worthington might be a good man. I also worked on Max Nicholson, Director of the Nature Conservancy. It took some persuading to convince him he ought to let his right-hand man go over to IBP, but he came round fairly soon. At a meeting of the Royal Society IBP Committee in June 1963, I suggested that Worthington be appointed Scientific Secretary of that committee, as a first step towards an international appointment. Then, when I went abroad at the beginning of July to a number of conferences and for a holiday, I asked Worthington to take over, unofficially, all the duties with respect to IBP which had come to me in my capacity as President of IUBS. These various actions were sufficient to get him started into the top full-time job of running IBP. He was officially appointed at the Paris IBP Assembly in July 1964, when Jean Baer took over the Presidency of SCIBP.

The early history

The seed sown by the 'founding fathers' as recounted above fell on fertile soil, for ecology in the late fifties and early sixties was emerging from a descriptive to an experimental stage. Production studies, growth analysis, photosynthesis, the flow of energy through ecosystems, were becoming subjects of research on both sides of the Atlantic, of the Eastern Bloc and also in Japan and Australasia. Prominent ecologists were communicating with each other and this made the time ripe for a co-ordinated international effort. In aquatic research, both freshwater and marine, production and population biology was in some respects more advanced and better organised than on land, perhaps because the industry of fisheries depends more directly on an understanding of natural ecological processes than does the industry of agriculture. In terrestrial ecology two leading figures in Europe were H. Ellenberg and P. Duvigneaud of Belgium, each of whom came to head their respective national IBP committees. A meeting in Brussels in September 1963, organised by Duvigneaud with strong support from the Belgian Government, marked the beginning of real activity by IBP's planners on terrestrial productivity.

The Evolution of IBP

To indicate how the programme began to get off the ground, two of IBP's International Conveners, the only two who were present at the early planning meetings and remained in charge of their sections throughout the decade of IBP, have written their personal accounts of the formative years. Although concerned primarily with their own sections, they provide some further points concerning IBP as a whole.

Conservation by E. M. Nicholson

As has been well brought out by C. H. Waddington, one of the early concepts for the entire programme had been to explore the realm of nature conservation, but this subject, like other early proposals, found insufficient favour and came to be superseded by a preference for biological productivity, linked with human welfare. It was at this stage that Jean Baer, then President of IUCN, invited the planning committee for the future IBP to meet in May 1962 at its new headquarters in Morges on Lake Geneva. By chance, I was there the same day on other business and was invited to sit in, together with a close colleague, Edward Graham of the US Soil Conservation Service, the moving spirit in IUCN's Commission on Ecology. Up to that point I had never heard of IBP, but after somewhat desultory discussion, someone rather suddenly proposed a series of seven working groups to plan the programme in detail, one being on conservation of terrestrial communities, which I was asked there and then to organise. After some reluctance I agreed, in view of Graham's generous offer to serve as my deputy.

A number of fundamental issues had to be faced. The mandate for the conservation section (CT) extended only over the terrestrial sector, the freshwater and marine sections being responsible for conservation within their own subjects. Moreover, inclusion of conservation in the programme did little to overcome the distaste felt for it by many biologists at that time. Few were attracted by its pretentions as a leading form of applied ecology, a subject on which also many biologists still cast a cold eye. The idea of harnessing conservation to the fulfillment of ecological research, and conversely of underpinning ecological research through a worldwide network of scientific reserves to be used as outdoor laboratories, was keenly held by Graham and myself, but its effective backing among both conservationists and ecologists was minimal at that time. Geographers, pedologists, foresters and zoologists displayed little more enthusiasm. Indeed, but for the support of the President of IUCN, and the possibility of exploiting a number of IUCN's worldwide contacts, it might well have proved necessary to report that a CT section in IBP was not viable. The difficulties were aggravated by the emphasis – inspired as it eventually proved – upon biological productivity, which meant little at that stage to the biological world, and set problems in gearing CT to its two terrestrial partners. However, a request to H. Ellenberg, who had been nominated as Convener of the section for terrestrial productivity, for an international basis of inventory or classification of vegetation types gained an immediate response.

The early demise of the original seventh section on education and training came as a blow, for high hopes had been set on such a body being able to stimulate academic institutions and others to provide courses geared to the various sections of IBP, not least CT, in time to yield a flow of able and vigorous recruits for teamwork in the field. The lack of such provision had a crippling effect, especially on the organising of expeditions to help countries with insufficient specialists of their own.

Perhaps the worst blow however, was the disappointment of the original expectations that IBP would enjoy a level of funding comparable to that of the preceding IGY. Years after IGY's official demise, I found, on a visit to Antarctica, expenditure there, largely in support of continuing IGY projects, running at an estimated $40 000 000 annually, while IBP was struggling along worldwide on a far less figure. Any appraisal of the results must take full account of this enforced poverty and its demoralising effect on many potential collaborators. CT could work through established conservation bodies in certain countries – especially the Nature Conservancy in Great Britain which provided many facilities free of charge – but even where this could be done it narrowed the range of biological contacts, and confirmed the idea in some quarters that IBP could be carried through on the cheap.

While the concept of including a special conservation section in IBP was theoretically valid, and the effort to implement it has produced results of some significance, it is open to debate whether the degree of communication and co-ordination with other sections has justified the original decision and whether in fact, had it been possible to launch an independent international project working within its own frame of reference to develop a scientific basis for worldwide conservation, better results might not have been obtained, and more ample support attracted. To pursue this argument would be unprofitable, and whatever might be the verdict today, the future biologists may come to appreciate and pursue lines resulting from the pioneer work of CT to which contemporary research workers were somewhat unresponsive. It must suffice for the record to point out that while certain elements in the IBP, and above all biological productivity, achieved gratifying recognition and support in the end, the biological community as a whole never fully endorsed in practise the inclusion of conservation.

Human Adaptability by J. S. Weiner

It was a most significant event for the human sciences to be involved in the IBP, because, during the 1950s and early 1960s, ecology as a discipline encompassing plant and animal communities was reasonably well established, but the study of human populations on a comparable scale lagged behind. Human ecology lacked both a body of field work and a theoretical framework, though in particular sectors important advances had been registered. This was the case, for example, in human autecology (the study of individual physiological adaptations), in some aspects of medical ecology such as malarial epidemiology, and of population genetics. But in setting up separate sections to deal with animal and plant communities the preparatory committee for the IBP was able to call on the various scientific unions already grouped within IUBS as well as on the IUCN, whereas ICSU contained at that time no organised body concerned with human biology.* Nevertheless, it was decided in 1959 that a worldwide stocktaking and analysis of the biosphere would not make sense if human communities were omitted. This view had the full support of the International Union of Physiological Sciences (IUPS) under the successive Presidencies of Sir Lindor Brown (UK) and Wallace Fenn (USA).

However, while the first promoters of IBP saw during the years 1959–62 quite clearly that it was to be based essentially on the ecology of terrestrial, marine and freshwater communities, the purpose and content of a human ecological component eluded them. The first plan called for a worldwide study of mutation and mutation rates in human populations, but this seemed too narrow to Sir Lindor Brown who, in consultation with me at Oxford in April 1962, accepted the view that an ecological programme concerned with human physiological, developmental, morphological and genetic adaptability was feasible and timely, and would form a fitting counterpart to the biological proposals. A programme along these lines was accordingly submitted to the Planning Committee at Morges in May 1962, which I attended on behalf of IUPS.

* As an outcome of IBP the International Association of Human Biologists (IAHB) was launched in 1969 and was affiliated to IUBS in 1973.

The IBP came just at a time when it could accelerate and catalyse the development of human population biology, which had begun already to supersede and transform the somewhat static subject of physical anthropology. The IBP required that living human populations had to be investigated as functioning entities interacting with their habitats, and therefore had to be understood in adaptive and selective terms. Accordingly Human Adaptability (HA) was adopted as the section's title.

From the beginning, HA laid stress on the comparative study of human populations in a very wide variety of habitats, and this comparative approach had a two-fold significance. Scientifically it represented an important way of discovering how far similar environments call for similar responses from populations of different genetic constitutions, and how far differing habitats may require differing responses from groups of similar genotypes. The issue was therefore to discover how far variations in genetic constitution, or in long-term exposure, imposed limitations and modifications by phenotypic response. The studies aimed at elucidating the interaction of nature and nurture on the physiological, morphological and developmental characters of human populations.

Much of IBP expressed a strong concern with 'natural' ecosystems subject to minimal interference for comparison with ecosystems which are subject to much change. The HA section in an analogous way placed emphasis on the simple societies living under difficult conditions. Such societies would provide object lessons of the actual adaptability achieved by man when relying on his biological endowment. The belief that IBP represented probably the last chance of making a concerted study of surviving communities of hunters and gatherers and simple agriculturists, proved to be justified even within the decade of the programme, for those people, living their traditional life, yielded information on the biological responsiveness of man to situations which soon will remain only in the folk memory of mankind.

The Morges meeting of May 1962 marked the beginning of a viable scheme for IBP. With little ceremony the Convenor appointed to each section found himself in charge of a biospheric portfolio of unimaginable magnitude, and the detailed planning during the following two years until the first General Assembly of IBP in July 1964 was crucial. In formulating realistic proposals for field research, acceptable to the scientific community, we faced the special difficulty of seeking out and communicating with human biologists of like outlook; but with the calling of the first HA conference at the CIBA Foundation in

December 1962, a network of interested and co-operative scientists emerged and began to grow. This multidisciplinary as well as multi-national gathering lasted precisely for the duration of the last great London fog, a circumstance which had the virtue of forcing a monastic seclusion on the participants who were thus obliged to remain together almost continuously for five days. The outcome was a clear framework and policy for the section with human adaptability as the central focus.

As a follow-up there was held, in July 1964, a scientific symposium under the auspices of the Wenner-Gren Foundation at their Burg Wartenstein Conference Centre in Austria. Twenty human biologists were able to spend a strenuous but profitable fortnight examining the aims of IBP in the light of their first-hand knowledge of particular communities and regions of the world (see Baker & Weiner, 1966). In consequence, the first version of the HA proposals for research, demonstrating what was at stake, was available at the General Assembly of IBP in Paris.

Then, during the discussions, there appeared in the midst of the delegates the formidable figure of Margaret Mead. The members listened with deep interest to her long and eloquent plea for the scrapping of the HA plans and the substitution of a programme based on the social sciences! This would have been a remit quite outside the overall proposals of the IBP and well beyond the province of the human biologists present. This incident has significance, however, because it symbolised the uncertainty which the scientific establishment in the USA felt at that time over the aims of the IBP as a whole. These aims, not merely of HA, took longer to command sympathy and support there than in most other countries, and one element in the USA attitude was undoubtedly the feeling that IBP was at fault in not coming to grips with the problem of overpopulation. But ICSU would have had to launch a very different programme, one perhaps more of a political than a scientific nature, if world population had been a central issue. In the event population and demographic issues found an important place in the HA work, and indeed could not be separated from the requirements of community research, but the scale of its consideration was limited by the constraints inherent in field research.

The proposals for HA which had emerged from Burg Wartenstein were accepted, almost with complete agreement, by the African, American, Asian, Australian and European human biologists present at Paris, so that, by the end of the First General Assembly, all concerned knew what was intended.

2. Substance of the programme

As recorded in Chapter 1, IBP had a five years' gestation period from 1959 before it was formally presented to the world at its First General Assembly in July 1964. Since then, it has held about 200 international meetings in many different countries; it has published a series of twenty-four IBP handbooks mainly on methods to be followed in carrying out the programme, and twenty-five numbers of *IBP News*; through its national or international agencies, some seventy volumes containing the results of symposia and meetings have been published; national programmes have been established and carried out in forty-four countries and fifty-three additional countries have participated in parts of the programme; more than 2000 separate projects, large and small, have been listed as contributing to the programme; some thousands of scientific papers containing results of these projects have been published, as well as numerous reports, reviews and separate volumes.

Those biologists who have led different parts of the IBP from the beginning, or who have emerged as leaders during its progress, together with new people who have examined the results, are now engaged in pulling all this material together in the form of syntheses. Many volumes of national synthesis which summarise and point up the conclusions of national projects carried out by each country have already been published and more are in preparation or are contemplated. At a different level is the series of international synthesis volumes to which this book provides a general introduction.

To understand why and how all this activity has come about, we must go back to the programme as launched in 1964, approved at the First General Assembly, and subsequently modified as experience was gained. This programme has already been published *in extenso* in *IBP News*, no. 1, revised in no. 9 and in certain other publications. In the formative years it was written somewhat unevenly, so this chapter is devoted to a restatement of the programme, reduced in length to about one quarter of the original. Some parts now read a bit 'dated', for the subjects concerned have moved fast during the past ten years; but it is against this programme that the successes and failures of IBP must be measured.

Before the programme was accepted there had been a number of discussions in working groups which came to form the nucleus of the seven sectional committees. Opinions ranged from those which would limit activities to a few carefully defined and circumscribed problems to be examined by standardised methods in as many places as possible, to those which would loosely define many aspects of biological productivity and human adaptability, from which each participating country could select those items in which its biologists were interested. There were good reasons for both these views and many intermediates. In the end the idea of a broadly based programme dominated, for environmental biology had not yet reached the stage of geophysics and climatology in which the variables worth measuring and observing could be defined precisely. Whereas different sections continued to take slightly different views, most of them set their goals wide, and in some areas of proposed research did little more than suggest general topics. By thus embracing a wide field, the programme presented a choice to participants in all parts of the world to formulate projects which they felt to be important and urgent and which they themselves felt capable of undertaking.

Emerging from these discussions, the subject of IBP was defined as 'The Biological Basis of Productivity and Human Welfare'. Its objective was to ensure the worldwide study of (*a*) organic production on the land, in fresh waters, and in the seas, and the potentialities and uses of new as well as of existing natural resources, and (*b*) human adaptability to changing conditions. The programme did not range through the entire field of biology but was limited to the basic studies related to biological productivity and human welfare, which were calculated to benefit from international collaboration, and were urgent because of the rapid rate of the changes taking place in all environments throughout the world.

General principles

Several common principles which applied to all parts of IBP were recognised as follows:

The fundamental approach. Although the simplest way to study resources is to carry out investigations on the resources themselves, it is also of great importance to make fundamental studies of the processes involved in their production. Therefore the programme set out to obtain internationally comparable observations of basic biological parameters.

Urgency. There was an urgent need to bring all parts of the programme into operation, not only because of the steadily growing pressures of human population on renewable resources but also because many of the situations, both biological and human, were changing rapidly, and many would soon no longer be available for scientific investigation.

Methods. There was a widespread need for elaboration and intercalibration of suitable methods. The series of IBP Handbooks which was designed to fulfil this need was expected to be of long-term value in itself, quite apart from its particular use during the programme.

Co-ordination and stimulation of existing research. Since much current research fitted into the programme, the function of the organisers was to co-ordinate rather than to direct. IBP benefited from all contributions relating to biological productivity and human adaptability, provided these were of good standard. By the judicious addition of a relatively few well trained workers or of appropriate equipment, IBP was able to stimulate research in a number of fields, and this was sometimes highly rewarding in relation to the expenditure incurred. Indeed, in some branches a whole new generation of biologists sprung into being.

Training. The programme needed trained research workers, some as leaders of IBP teams, others in the use and application of recommended methods. A number of short-term courses and workshops in particular subjects were planned, both nationally and internationally, for example in bioenergetics and systems analysis. In the event, however, most of the new biologists learned their trades in the field rather than in formal courses of study.

International co-ordination. The plan for IBP was in some cases greatly helped by the advice and support from a number of other international organisations, both intergovernmental and non-governmental. In order to avoid unnecessary duplication, collaboration was arranged, for example with the Intergovernmental Oceanographic Commission, the International Hydrological Decade, the study of man-made lakes, as well as the programmes of sister committees of ICSU, such as SCOR, SCAR and COWAR.

Storage and retrieval of data. Some parts of the programme would involve collecting quantities of data of a kind which did not lend themselves to immediate publication, but might have future importance. Special facilities for the storage and retrieval of such data were considered highly desirable, but in the event proved difficult to arrange in most cases.

Publication of results. All primary publication of IBP research would be arranged through existing international or national channels, such as scientific and technical journals and book publishers. Only in the case of supplementary publications such as reports of important meetings,

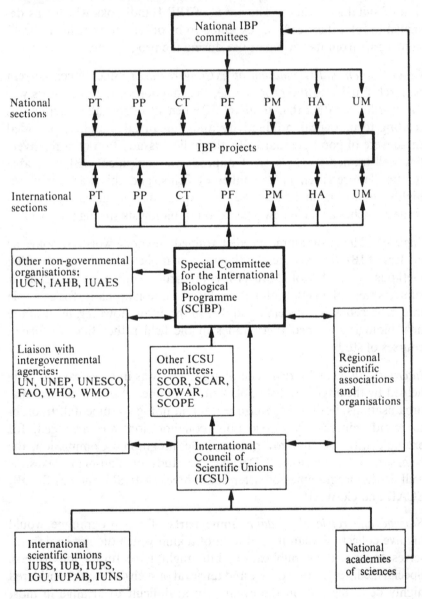

Fig. 1. The organisational structure of IBP.

handbooks of methods and synthesis of results, would special series of publications be contemplated.

With these principles in mind, the programme was arranged in seven sections, three dealing mainly with terrestrial situations, two with aquatic, one with the human species, and the last with use and management of resources.

The *sections* were to a considerable extent autonomous in their organisation, but broadly each included a number of more or less distinct subjects which were referred to as *themes*, and each theme was composed of a number of more or less related *projects*. Each project was planned, staffed, financed and directed nationally by a participating country, or in certain cases by two or more countries working in concert. Thus the international programme for each section was carried out by a number of national projects within that section, each participating country focussing its work on those themes in which it had particular interest or expertise.

The content of each section's programme is summarised in the rest of this chapter. This is in no sense a report on what has been done, but a statement of intent at the commencement. Therefore it is written in the future tense.

Productivity Terrestrial (PT)

Many kinds of study fall within the sphere of PT, but biological production at different trophic levels will provide a unifying factor and should be a fundamental part of every project. Descriptive surveys will often be preliminary to productivity studies, but are not an end in themselves. Two kinds of project can be recognised: (*a*) a few major, really comprehensive, projects of which the aims are to estimate the primary and secondary production of a particular site, and to establish its overall energy flow; and (*b*) supporting projects arriving at the evaluation of production or energy flow of certain types of plant communities or of a given level of consumers.

As far as possible, studies of primary and secondary production should be undertaken in the same sites. Projects should aim at comparing biological productivity of wild and man-modified ecosystems in the same environmental conditions. There should be comparative studies of the same type of widespread organism in different environments (e.g. *Phragmites*, *Pinus*, bracken; wild and domestic herbivores, granivorous birds, grasshoppers).

Selection of sites and communities

The siting of projects should reflect a particular development in the range of gradient of a plant community. Sites should permit comparisons of different communities on areas of similar soil, topography and climate, in order to test the degree to which different biotas have evolved comparable efficiences in the production and utilisation of organic matter. Additional sites should reflect the effects of such variables as soil, topographic position and exposure, community physiognomy and age of stand. Other sites may reveal how the 'natural' production can be increased through the application of agricultural and silvicultural techniques. Thus the sites studied will vary from: (i) natural, through (ii) a range of man-modified communities, to (iii) experimental and other highly artificial systems.

The sites little modified by man, which owe their productivity to natural selection, provide a base against which to evaluate the effects of human activities. Particular attention should be paid to determining the proportion of the total biological production currently harvested by man.

The communities should be selected to be representative of significant major ecosystems and to take advantage of special situations, such as dated lava flows or land exposed by glacial retreat, which permit precise time studies. Similarly, it is possible in certain regions to compare adjacent natural and artificial communities on the same site or to study plant species important for human welfare over their geographic range. Cereals are important in this respect and trees of restricted natural range but widely planted. Of importance are communities that are considered most productive (e.g. artificial grasslands, well manured field crops or tree plantations with dense canopy), since it would be of value to know the maximum primary production being obtained in the principal regions of the world. Descriptions should specify community composition, both taxonomically and ecologically, in order to permit valid comparison between sites. At each field site, vertical stratification and horizontal variability of the community should be shown by transects and maps.

Primary production

Investigations on the production of plant organic matter by photosynthesis and chemosynthesis should be made at a global network of sites, sufficiently well distributed to characterise the overall system. They should be based on agreed methods and should take into account the dynamic nature of plant/animal communities. A starting point is the net dry matter production of green plants over the year or vegetation period. This can be determined by the sum total of (i) biomass change, (ii) plant losses by death and the shedding of parts, (iii) consumption of plants by animals, and (iv) man's harvest where applicable. The amount and area of leaves should be measured in each community, since a leaf area index permits comparison of results. These studies could be supplemented by chlorophyll investigations.

At some sites additional effort should be concentrated on studying the detailed functioning of ecosystems and here it is essential that primary and secondary production investigations should be combined. Research on primary production at these sites may involve studies of energy flow and the movement of materials through ecosystems, respiration, pollution and decomposition, as determined with the aid of elaborate equipment. Measurements of chemosynthesis are envisaged where relevant.

Secondary production and decomposers

Where possible, secondary production should be studied on sites where primary production is also being investigated. It should be concerned particularly with energy flow and the circulation of materials within ecosystems. There are three main aspects of secondary production: (i) those directly related to man's current and future needs, concerned, for example, with domesticated and wild herbivorous mammals, (ii) those concerned with animals which can bring about decreases or increases in primary production, for example the feeding activities of many invertebrates, birds and small mammals, and (iii) those processes occurring within the soil and litter layers which are concerned in the circulation of material.

It is necessary to determine not only the consumption of vegetation but also the predators, sometimes through several links of a food chain, and parasites. This will necessitate faunistic surveys in which emphasis is directed towards determining the utilisation of net primary production.

The conversion of material through the grazing food chain is more easily determined than through the detritus food chain; though in some circumstances adequate estimates of the latter can be obtained by difference. Two basic approaches are possible for estimating the utilisation by consumers: estimation of removal rate, and estimation of the

23

respiration of the standing crop of grazers and, knowing the assimilation efficiency, calculation of the food requirements as a percentage of total net primary production. Measurements beyond these first steps are difficult, especially in detritus food. Such studies will have to be restricted to a few centres where there is close integration between botanists, soil microbiologists, zoologists and pedologists.

In most situations the field ecologists will have to determine the numbers and biomass of dominant organisms and the relations of these organisms to the trophic levels in which they occur. The age structure of populations should be determined from life-table data or by some other suitable procedure to give estimates of turnover. Physiological indices (food intake and utilisation, respiratory rates and energy conversion factors) will be necessary to convert basic ecological data into production values.

Concerning the decomposer cycle, soil organisms, in particular bacteria, fungi and actinomycetes, play a most important role in the transformation of animal and plant residues and in their final return to the soil on complete mineralisation. The micro-organisms control such cycling as that of carbon or nitrogen and sometimes may inhibit the normal processes of humus formation.

The qualitative and quantitative studies of the numbers of soil organisms and the measurement of the intensity of their various activities (for example decomposition of cellulose, lignin, pectin and processes such as ammonification) are some of the most important problems. Whenever possible, the measurements should be designed to provide evidence which will help in understanding the quantitative changes associated with the general cycling of carbon, including the proportions of carbon incorporated in humus and in carbon dioxide loss from the soil. The comparative study of the formation of humus, and its nature under different climates and in the principal ecosystems corresponding to these climates, is of particular importance.

Environment in relation to production

Existing meteorological networks should be used so far as possible to relate production to regional climate. However, the relation between plant productivity and climate depends on the interplay of many complex physical and physiological processes and there is a need to distinguish those climatic elements of the greatest biological significance. Owing to the importance of energy for photosynthesis and transpiration, special attention should be paid to the measurement of solar and net radiation. Other relevant variables are air temperature, humidity, wind and precipitation. Microclimatological studies need profiles of temperature, humidity, wind and carbon dioxide from the soil surface upwards.

The aid of pedologists must be sought for the description of soil profiles and to relate these to the accepted soil maps and types. Soil texture, structure, total volume, humus content, pH, exchange capacity

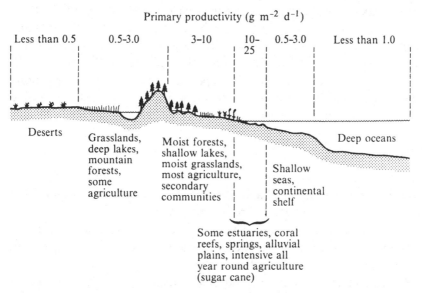

Fig. 2. An early assessment of global primary productivity. This diagram, based on one published by H. T. Odum in 1959, provides a perspective for the biome and ecosystem studies of IBP which will elaborate and add significance to the estimates of production.

and chemical composition for the horizons used by plants, should be recorded at every station. Soil water distribution during the growing season and the wettest and driest parts of the year should be evaluated and supplemented, where possible, by comparisons of actual water supply and use, with potential evapotranspiration graphed through the years. Past or present interference by man must be assayed carefully in each research area.

Ecosystem processes

The study of primary production, of diverse forms of secondary production, and of environmental factors, permits a better understanding

25

of the functions of ecosystems at different trophic levels. This under-standing must be based on knowledge of organic production and turnover, energy flow and the circulation of water, carbon and other essential biogeochemical elements. Chemical analysis of elements including N, P, K, Ca, Mg, Mn, Fe, B, S, Mo, in plant, animal and soil material is necessary to discover factors controlling production and to measure the quality of the organic matter present. Energy determinations by calorimetry will also be needed and problems of nitrification and nitrogen fixation should be studied in collaboration with Section PP.

The geographical variability of the biochemical activity of plants and animals may permit developments in the economy of some countries. Importance is attached to the discovery of plants, wild or cultivated, capable of maximum photosynthesis per unit area of foliage, or ground surface, or capable of maximum production of the proteins or amino acids necessary in the human diet.

General co-ordination

Results should be expressed in agreed metric units and productivity data given on an area–time basis. Precision, reproducibility and sampling error should be estimated throughout. Particularly during the first phase, emphasis should be given to the development and testing of methods and to the interpretation of results so that valid comparisons can be made between different ecosystems.

Related international and national research already functioning should be supported or incorporated whenever possible. Among those in the international sphere are projects of the IUBS Committee for Biological Control, the UNESCO/FAO project of mapping world vegetation, the FAO/UNESCO/WMO Working Group on agroclimat-ology, the UNESCO Arid Zones and Humid Tropics programmes and the International Hydrological Decade. In the Antarctic, close liaison with SCAR is essential. In co-operation with Section CT all research sites should be protected, at least during the period of investigation, against developments which would damage their scientific value.

Production Processes (PP)

Two processes having a key position in productivity of ecosystems should especially be studied, namely the use of solar energy by plants in photosynthesis and the fixation of molecular nitrogen by living organisms. Greater knowledge of an improved methodology for evaluating

26

these processes can be important in the development of more productive natural and artificial communities. Comparative investigations are essential and should include different conditions for natural, agricultural, and experimental systems.

Biological fixation of nitrogen

The field of activity was originally designed to include both biological fixation of molecular nitrogen and transformation of nitrogenous compounds. Although the latter represents an important stage of the nitrogen cycle, activity should concentrate on nitrogen fixation. This involves the study of ecological, physiological, biochemical and agronomical aspects, and it comprises many different kinds of system. The prime objective is to study the principles rather than the practical applications, and especially those factors which determine the highest possible productivity under particular environmental conditions.

Biological nitrogen fixation is caused by free-living micro-organisms as well as by symbiotic systems. Of the free-living organisms two groups are concerned:

(i) *Bacteria* – although in general it is believed that free-living nitrogen-fixing bacteria do not contribute to a large extent to the nitrogen conten of the soil, there are reports on particular sites containing large numbers of nitrogen fixers. A worldwide survey of such habitats should be carried out including qualitative as well as quantitative microbiological investigation. A detailed description of soil character, climate and vegetation will be necessary for each habitat to see which factors govern this type of nitrogen fixation.

(ii) *Blue-green algae* are thought to fix considerable amounts of molecular nitrogen when occurring under aquatic conditions or in certain types of wet soils such as rice fields or soils with an undisturbed surface layer. Knowledge of the distribution of these organisms in nature is restricted and therefore a survey will be made using the heterocysts technique. This method uses the number of heterocysts (thick-walled, colourless, empty-looking cells) as a measure of the nitrogen-fixing potential.

Symbiotic nitrogen fixation provides by far the largest amount of nitrogen to higher plants. This applies both to the associations which involve leguminous plants and bacteria of the genus *Rhizobium* and the associations which involve non-leguminous plants and bacteria of the actinomycete type. In both kinds of symbiosis, efficiency depends on a number of internal and external factors. The former include the species

27

and variety of plant and the strain of *Rhizobium*; the latter include climatic and soil conditions. Very often not all of these factors are optimal. Study of their influence on nitrogen fixation and plant growth should be carried out as follows:

(i) *Leguminous plants.* A comprehensive survey of plant–*Rhizobium* systems in order to identify those which fix high amounts of nitrogen would be highly desirable, but impossible within the period of IBP. Therefore a less ambitious programme is planned, namely a study of symbiotic nitrogen fixation by way of three standard experiments consisting of a field trial, a glasshouse trial, and a soil-dilution experiment. The purpose of the first two is to estimate the nitrogen-fixing capacities of suitable legumes (such as lucerne, soybean) in representative areas. Inoculation with an effective *Rhizobium* strain, and soil improvement with lime, fertilisers, and combined nitrogen, are introduced as variables. The soil-dilution experiment is for obtaining a quantitative estimate of numbers of nodule bacteria in the soil.

A second study will consist of the search for symbiotic systems adapted to less favourable climatic or soil conditions. It includes such factors as high and low temperatures, and the rooting medium, soil acidity and molybdenum deficiency. Such adapted systems may be obtained by selection or breeding of particular varieties of plants, or by selection of particular strains of rhizobia.

A third study will result in a world catalogue of rhizobial strains at present available in national collections.

(ii) *Non-leguminous nitrogen-fixing plants.* Although the number of such plants is much smaller than that of the legumes, many species are known to occur and others are suspected. A survey of certain genera will involve the inspection of a large number of plant species for root nodules.

In addition to root nodules, some non-leguminous plants bear leaf nodules in which it seems that the bacteria, in addition to fixing nitrogen, provide the plant with growth substances. Since leaf-nodule-bearing plants may be found in the same regions as the root-nodule-bearing non-legumes the proposed survey will cover both kinds.

Photosynthesis and solar energy conversion

The scope is defined as the fundamental physiological analysis of processes governing photosynthetic production per unit area. Research at

28

all levels of structure: community, individual, organ, cell, and at all degrees of sophistication relevant to this aim, is included. The photosynthetic production of experimental, agri-, sylvi- and horticultural and natural communities and their components will be investigated. Included in this research will be:

(i) *Experiments requiring only simple equipment,* in which the dry matter production and energy conversion in photosynthetic systems is investigated using the techniques of growth analysis. Among these will be studies on the effects of climatic condition, plant density, irrigation, fertiliser treatment and light intensity.

(ii) *Experiments requiring elaborate equipment and personnel.* These comprise measurements of the photosynthetic production of plant communities and their components. They will relate gas exchange and solar energy conversion to the environmental factors, energy budget, stand structure and physiological status of the plants.

The elaborate experiments will involve fundamental investigations of the limitations to photosynthetic activity at various levels (community, plant, organ, cell) by environmental, physiological and genetic factors; and studies of photosynthetic productivity as affected by morphogenetic reactions of plants to the environmental factors. Both the simple and elaborate experiments involve comparative investigations of photosynthetic production in specific types of plants, including trees and shrubs, perennial herbs, annual and perennial crops, marsh and aquatic higher plants, mosses, freshwater and marine algae, and also in communities and cultures under various environmental conditions. Elaboration of comprehensive mathematical models of photosynthetic production and growth in plants and plant stands will also be included.

Other biological processes

While the main focus will be on the nitrogen cycle and photosynthesis, other processes of importance to biological productivity should be included in collaboration with other sections. For example:

Phosphate metabolism. The main factors which limit the efficiency of biological production include water supply, temperature, light intensity, CO_2 tension, and mineral deficiencies. Among the mineral deficiencies phosphate is of major importance. Comparative studies on the role of phosphate in primary production are therefore envisaged.

Protein production by plants is a consequence of photosynthesis, and may proceed very rapidly: amino acids can be traced within seconds after feeding $^{14}CO_2$. In the absence of adequate nitrogen, carbohydrates can accumulate during a light period and link up with the nitrogen later. As with other assimilates, studies of protein formation may be carried out with higher plants and also with algal cultures.

Secondary production is of the greatest importance to productivity but is left mainly to other sections. However, the links between primary and secondary production processes, e.g. in problems of efficiency of energy conversion of mechanisms of energy-rich phosphate formations, of mutual relations between autotrophs and heterotrophs, are so important that close collaboration will be maintained.

Conservation Terrestrial (CT)

One of the first tasks is the examination of the range of ecosystems over the world and the assessment of the extent to which scientifically adequate samples of all the main types and their significant variants are already protected as national parks, reserves or research areas. Since it is well known that in many regions the existing series of protected areas is inadequate, the assessment will also have as a main aim the selection of other areas which should be preserved. How to provide for the protection of many scientifically interesting species whose survival cannot be ensured solely by the setting aside of reserve areas will also be considered.

The preservation of natural and semi-natural areas is important for the future of biology and for human welfare because they provide for:

(*a*) the maintenance of large, heterogeneous gene pools;
(*b*) the perpetuation of samples of the full diversity of the world's plant and animal communities in outdoor laboratories for a wide variety of research;
(*c*) the protection of samples of natural and semi-natural ecosystems for comparisons with managed, utilised, and artificial ecosystems;
(*d*) outdoor museums and areas for study, especially in ecology and;
(*e*) education in the understanding and enjoyment of the natural environment and for the intellectual and aesthetic satisfaction of mankind.

In addition to the selection of a series of reserves, the CT section seeks to ensure a scientific foundation for conservation throughout the world, and in particular for the assessment of conservation problems, for the management of plant and animal communities and species, and for training of specialists and the education of a wider public in the aims and practice of scientific conservation.

While the section will concentrate primarily upon the land, contact with the PF and PM sections will ensure that specialist knowledge of freshwater and marine biologists is associated with that of terrestrial conservationists in a comprehensive effort.

Selection and survey of areas

A final list of recognised and proposed reserves should include sites which, although they may not be essential to illustrate the range of ecosystems, support species of plants and animals of outstanding interest or great rarity; sites which are of scientific interest because of the human management to which they have been subjected, even if this has in some cases led to more or less far-reaching modification of the biota; sites which are important because they have been the scene of detailed and well documented research; sites which contain, for example, deposits of peat, lignite or sediment from which information may be obtained about past vegetational and climatic changes, and also sites which are of special palaeontological importance; sites which are of special physiographic or geomorphological interest and which represent unusual habitats.

The following general considerations will also be borne in mind:

(*a*) Sites should be included whether or not they appear to be under immediate threat.

(*b*) Preference should be given to sites than can conveniently be worked by existing or proposed research institutions; to sites that can be supervised and managed effectively; and to sites least likely to be affected by adverse development or pollution.

(*c*) Areas must be of adequate size to support viable populations of the species which characterise them.

(*d*) Research areas must also be large enough to allow for the increasing amount of land demanded by modern field experimental research.

(*e*) There are sometimes advantages in locating scientific reserves near to or within larger areas of high landscape value, since it then becomes justifiable to protect a larger and more viable unit.

Two main processes will be involved. Firstly, an inventory will be prepared of the major world ecosystems using vegetation formation in combination with a variety of other data. This will be amplified by compilation of national inventories of the significant ecosystems and variants occurring in each of the participating countries.

Secondly, a standard Check Sheet will be used to record information in a rigorous and comprehensive manner about a whole series of areas which may merit consideration for protection. These areas may be either all or part of an existing national park, nature reserve or other safeguarded area, or some sample of an unprotected area which possesses scientific interest. In order to handle and make full use of the large amount of information expected from the check sheets, an international data centre will be established with appropriate equipment and staff.

The preparation of national inventories, and the completion of check sheets, should normally be part of national programmes. Where this cannot be done, information will be sought either from individual scientists in the countries concerned, or from foreign scientists with special knowledge of those countries, or through expeditions.

The study of areas in depth

An important element is the improvement of knowledge of scientifically important areas through detailed ecological surveys and research. In this the detailed investigations of other sections will enhance knowledge of many reserve areas in a variety of ways. Specialised ancillary check sheets to obtain supplementary data, such as conservation status and research facilities will be used on selected areas.

Assessment of conservation requirements for rare and threatened species

Conservation of species is accomplished primarily by the effective protection of the habitat. However, certain species may not occur within, or may not be limited to reserved areas, and many species of fauna regularly migrate. Learning the scientific requirements for the conservation of each species is therefore of great importance. The *IUCN Red Books* for endangered species of mammals and birds (Simon, 1966; Vincent, 1966) and the information held by the IUCN Survival Service Commission provide the basis. IBP's objective here is to provide the IUCN with additional scientific data including that from plants. The list of species can then be examined in relation to the Check Sheet Survey and results of research on movements, habitat requirements and population dynamics, and a judgement made as to the further measures required for their conservation.

Scientific basis for the protection of areas and species

As a result of IBP the extent and effectiveness of protection afforded to ecosystems, species and sites of outstanding scientific interest will be increased. The CT section is not directly concerned with the measures that this process requires, but with providing the scientific basis on which implementation in the form of enforcement, administration and legislation, must depend.

General investigations on conservation problems

Much new detailed scientific information must be obtained to facilitate the appropriate management of sites set aside for protection. This should be done concurrently with the surveys. Among investigations required are: studies of the dynamics of plant and animal communities, and of successional trends; studies of the autecology and population dynamics of individual forms of plants and animals, and especially of vulnerable, or particularly interesting species; studies of the manipulation of plant and animal communities, especially through the control of grazing by domestic or wild species, and the regulation of hydrological regimes, predation, and other factors; and studies of the impact of human activities on natural and semi-natural plant and animal communities, and hence the identification of those forms of human use that are wholly or in part incompatible with the objects of reserve management or the maintenance of a given ecosystem. Taken together with the surveys on existing and proposed reserve areas, and with wider ecological knowledge, these studies should lead to a clearer definition of the conservation problems in the various regions. They should also indicate for the benefit of national authorities the requirements for an adequate national conservation programme, and the research potential of the areas studied.

Productivity Freshwater (PF)

The information sought should point the way to methods for obtaining more food (by increasing productivity) and more clean water (by reducing productivity and pollution). Thus the objectives can be stated: to study the basic factors of production and metabolism, at all trophic levels, in standing and running waters, both natural and polluted. To do this in a co-ordinated manner on a global scale, substantial preparatory efforts are required. Methods have to be improved, research workers trained, and each project planned in detail. Particular care will be needed to achieve results from the individual projects that are comparable, and where possible, the projects should be selected to stimulate pre-existing research and thereby achieve maximum results with minimum expenditure.

Conservation of aquatic ecosystems – Project Aqua

In order to provide a baseline against which the productivity and energy turnover of freshwater communities can be measured, it is vital and urgent to conserve and maintain for permanent study examples of the different types of ecosystems, suitably distributed through the various climatic zones and biogeographical regions of the world. A list of examples of such ecosystems will be prepared in co-operation with IUCN and SIL, and will be co ordinated with similar work by Sections CT and PM.

Climatic zones

The programme should utilise existing research centres where appropriate facilities and scientific guidance are already available. In addition, special provision should be made for the study of important ecosystems far removed from research centres. In order to cover the range of communities of standing and running waters without dispersing the effort too widely, and bearing in mind that fundamental problems are often best solved in systems which are ecologically simple, a selection of centres for the programme should be made. Such centres should be sited strategically in the Arctic, north temperate, tropical, south temperate and Antarctic zones.

The Arctic and sub-Arctic regions are sources of high-grade fish, and the relative paucity of species in their biotic communities might enable a rapid understanding of the processes governing their productivity. Close contact should be maintained with the biological work of SCAR.

In the temperate zones, where a number of hydrobiological institutes already study productivity, the chief need is for the investigations to be co-ordinated with those being carried out elsewhere. For lakes and streams this could be effected by co-operation of existing institutions in Eurasia and America; but rivers need special emphasis. It is in the north temperate zone, where most freshwater scientists work, that new and promising methods of investigation could be developed, but attention should also be directed to the south temperate zone where it is desirable to increase the number of institutes.

In the tropics, where fresh waters tend to be of special importance in relation to the local economy and food supply, the programme should be based where possible on the few existing research stations, and also in some cases on sites of common interest in major catchment areas such as those of the Nile and Amazon. In this climatic zone a distinction can be drawn between the drier regions where freshwater systems, though often extensive, tend to accumulate salts, and the humid tropics where there is a large throughput of water of low salt content.

Phase I

This first phase will be concerned primarily with the following:

Consultation with hydrobiologists. The first report of the IBP Planning Committee on fresh waters, prepared under the guidance of Professor Rhode, was distributed for comment to the XV International Congress of Limnology in August 1962, and the second report (October 1963) was commented on by a substantial number of specialists throughout the world. Thus, the policy of wide consultation was started in advance.

Training of research workers. The world supply of biologists has in the past been barely sufficient to replace those retiring. This is particularly evident in the fields of taxonomy and limnology. It is important that extra provision be made for training in these fields, especially in the tropics. The establishment of studentships and fellowships will be necessary, and panels of experienced limnologists will be needed to advise research teams in less advanced regions. Existing institutions and organisations should provide training, space for visiting workers, and, if possible, secondment of staff for research elsewhere.

Methods of productivity research. Initially, methods will be examined to assess the chemical environment, primary, secondary and fish production.

Phase II

National programmes should endeavour to study each of the major trophic levels in at least one representative standing or running water in respect of community structure, biomass, rates of biomass changes, factors controlling biomass changes, and utilisation efficiencies. It is as important also to investigate historical processes as they are related to the changes of productivity. Special attention should be devoted to fishponds, particularly in the tropics.

Ordinarily, the smallest efficient unit to undertake a project would be a group of four or five scientists, plus assistants, boats and shore facilities. In many places a small addition of personnel and equipment and a slight modification of the existing programmes could achieve the required results.

The following specific topics have emerged as suitable for inclusion in national projects:

Radio-isotopes offer an excellent means of studying the flux of materials through the ecosystem. So far as is known, all freshwater organisms are capable of accumulating isotopes from their environments. In standing water the accumulation is liable to be more rapid and recirculation may be continuous, whereas flowing water is apt to carry away and disperse the materials.

Salmon fisheries in the Arctic and nothern temperate zones provide a bridge between freshwater and marine communities. [*Note:* In the event nothing was done on salmon by IBP, but the subject was developed with vigour by other agencies.]

Wetlands are extensive and little is known about them in many parts of the world. Their study, which should include economically important plants, forms a link with Sections PT and CT.

Transfer of material from land to fresh water has large effects on productivity. Special cases include material eroded from soils and the faeces of mammals and birds.

Backflow of energy from water to land includes, for instance, aquatic insects on reaching imago stage, and land animals feeding in water.

Transfer of material to the sea by rivers adds much fertility in the form of plant nutrients and organic matter. Study of this forms a link with Section PM.

Large impoundments have been made or are planned in many parts of the world. Hydrobiologists will be available to study several of these schemes, and IBP should co-ordinate and augment their work.

Silt in suspension influences productive rates and has particular importance in countries which are subject to major soil erosion.

Artificial heating of aquatic habitats through various industrial processes should be studied more closely.

Nitrogen fixation in freshwater ecosystems may explain some differences in productivity between bodies of water. This project involves co-ordination with Section PP.

Extra-cellular products in solution, produced by actively photosynthesising algae, make a contribution to productivity which has not yet been assessed.

Role of bacteria in the energy flow through aquatic ecosystems, especially in the return of nutrients to circulation, likewise needs study.

Metabolism in relation to age of organisms should be studied for evaluating productivity.

Toxic chemicals, through purposeful or accidental application, are known to have particularly marked effects on aquatic communities. These should be studied with special reference to productivity.

Detritus is a major source of nutrients and energy for secondary producers. In many running waters it is by far the most important source, yet quantitative aspects are little understood.

Organic substances: their formation, assimilation and decomposition by algae should be studied.

Palaeolimnology: the determination of the trophic history from organic sediments is of great interest.

Co-operation between biologists and mathematicians. When dealing with ecosystems larger than a pond it is usual to assume random distribution of many observed variables. As nature is in fact not random, attention should be given to this dilemma.

Flowing waters. Finally, it is stressed that much more is known at present about productivity and energy flow through ecosystems in lakes than in running water. Section PF should arrange a considerable amount of research on rivers and streams.

Productivity Marine (PM)

Improving understanding of the basic ecological mechanisms which control the abundance, distribution and production of marine organisms is of particular importance in inshore areas where the effects of man's interference are so great. It is these areas which offer the greatest prospect of improving resources through human intervention; they include the cultivation of marine invertebrates, fishes and seaweeds in which there is great capacity for development.

The detection and measurement of ecological mechanisms will depend in part on comparisons between different environments: coastal and oceanic, shelf and slope, tropical, subtropical and temperate. Long-term variation needs to be studied as this is a major source of difficulty in planning the efficient utilisation of marine resources in many parts of the world. There already exist a number of well developed international organisations in marine science, particularly in fishery biology, with which IBP must collaborate very closely. There are, however, opportunities for improving communication between marine scientists and this would be especially valuable to workers in developing countries.

The central programme will be the study of seasonal variation and, through international collaboration, the comparative analysis of spatial variation. Supporting programmes will be designed to ensure that the interpretation of variation is sustained by adequate knowledge of the physiology and ecology of the most important organisms and communities.

Central programme

The objective will be to identify and measure the organisms, detritus and solutions in which organic material is held and the rates at which it is synthesised and passed from one organism to another, coupled with the measurement of related physical and chemical factors. The recent rapid development of physical or chemical methods for assessing the organic content and rates of production in the sea gives the opportunity for intercalibration and provides short-cuts towards the understanding of biological production. The application of these new methods, and of some established ones, to a study of temporal and spatial variations would lead to a considerable advance of our knowledge of organic production.

All laboratories collaborating should operate a basic minimum scheme which the better-equipped institutions would incorporate into more ambitious studies depending on the available facilities and the special interests and skills of the scientific staff. Laboratories with small vessels would be restricted to inshore water, but the larger vessels could be used to extend the programme further out to sea and a special attempt could be made to study the differences between shelf and oceanic waters in the region of the continental slope. The foundation of the programme

38

would be the repetition of sampling throughout the year with special attention to periods of rapid change such as the spring outburst of phytoplankton in temperate regions. Each laboratory would design its sampling programme in the light of the local conditions and it is suggested that standard stations be defined in the form of traverses stretching from the shore to the seaward limit attainable.

The notes which follow, which illustrate the kind of observation needed, are biased towards the plankton, but there is equal need to improve knowledge of the part played by the benthos in the production of organic matter. The problems of sampling and estimating production and turnover in the benthos are however, formidable.

Hydrography. There is little difficulty in reaching agreement on the comparison of measurements of salinity and temperature. It is highly desirable that stability and mixed layer depths be studied as a background, for example, to the investigation of phytoplankton blooms and nutrient circulation.

Solar radiation. It is necessary to measure both incident and submarine illumination. Ocean weather ships, merchant vessels and space satellites might assist in this part of the programme.

Nutrients. New methods have been developed recently for the determination of reactive phosphorus and nitrate in seawater. Laboratories with suitable facilities and man-power might study other nutrients, particularly ammonia-nitrogen.

Dissolved organic matter and detritus. There are relatively large amounts of dissolved organic substances in seawater, but little is known about their temporal and spatial variations or of the part which they may play, together with organic detritus, in heterotrophic growth. There have been recent improvements in methods for the measurement and identification of organic materials in detritus, but, as with the problem of organic material in solution, there is as yet very little information.

Phytoplankton pigments. The measurement of chlorophyll *a* provides one of the best means of estimating the total plant material in the sea. The spectrophotometric method has become standard and recent work has led to improvements in methods of extracting the pigment, in calibration, and the equations used to express the results.

Rate of photosynthesis. The uptake of ^{14}C must be used to measure photosynthesis in oligotrophic zones though the simpler method of measuring oxygen production may be adequate in eutrophic zones.

Phytoplankton counts. An investigation should be made in order to develop a standard procedure based on the Utermohl technique for counting cells in water samples.

Zooplankton. Quantitative sampling of zooplankton is extremely difficult. It may be necessary to limit IBP work to a form of net which will provide an adequate sample of the majority of the herbivores; but there may be strong practical as well as scientific advantages in using a high-speed sampler.

Fish. In placing fish and other resource stocks including mammals, in the trophic network of the sea, it will be necessary to distinguish between those stocks which are used by man and those which are not. In the fished areas the catch per unit effort by statistical squares is available for a variety of species. Estimates of abundance should be checked by other methods, including egg and larval surveys. In the unexploited areas such surveys might be supplemented by acoustic techniques.

Biogeography. Studies of production would be of limited value without knowledge of the distribution and abundance of the most important species and communities. In some areas this knowledge exists; in others it will be necessary to solve fundamental problems of taxonomy and morphology. Special attention should be paid to fish eggs and larvae, fish food and predators, and the detection and plotting of unconventional resources as well as those which are exploited by man.

Supporting programmes

The kind of observations outlined above should be amenable to analysis by multivariate statistical techniques, and should provide some of the principal terms required for the formulation and testing of mathematical models of marine productivity. The weakness of such models is that they can be designed only from assumptions based on existing knowledge and the results can be interpreted only in terms of the selected parameters which may not be the controlling factors. It is essential, therefore, that projects in the central programme should be closely linked with studies of controlling mechanisms. A great variety of field and laboratory projects is required to examine such mechanisms, to fill gaps in knowledge, to develop new techniques. One major topic requiring collaboration is the estimation of productive rates in animal populations as a parallel to the estimation of photosynthetic production. Other examples are provided below.

The study of nitrogen as a primary nutrient should not be divorced from knowledge of nitrifying bacteria, involving work on populations in the sea as well as metabolic studies in the laboratory. Nutrients are estimated in the sea by chemical methods, but we need to know whether they are all usable by organisms; biological tests or assays may be necessary. Dissolved organic substances should be investigated in relation to particular organic material in living organisms and detritus. The chemistry of excretion by both plants and animals should be studied in this context.

There is a great deal of information about nutritional requirements and general physiology of a few organisms, such as *Calanus* and *Balanus*, which for various reasons have been studied extensively. However, we are ignorant about important organisms in many marine communities. Bacteria and flagellates require particular urgent attention as they are so numerous and their part in the circulation of organic material must be very great indeed. The blue-green algae and seaweeds are groups in which the ecologist needs the help of the physiologists. An important objective of experimental physiology should be the study of seasonal and other temporal changes. For example, it is known that growth factors are important in diatom development but very little is known of spatial or temporal variations in either the requirements or the abundance of these substances in the sea.

It is assumed that there will be full collaboration between physical and biological oceanographers, but there are special topics in which joint work is particularly important. For example, there is insufficient knowledge about internal waves in the sea and their effect on the distribution and abundance of organisms; such effects may be particularly important in considering the differences between shelf and slope regions.

Participation

The distribution of marine laboratories is such that there is likely to be adequate participation from the northern hemisphere. However, it is important for the achievement of the proposed comparative studies that tropical and sub-tropical programmes be included as well as programmes in the temperate parts of the southern hemisphere.

It is regarded as essential that individual workers and teams retain their own autonomy so that they can control and plan their own programmes and reap their own rewards. The success of the programme as a whole will depend on the free interchange of plans and information; it may be that one of the most useful functions of IBP will be to provide the opportunities for such interchange.

Human Adaptability (HA)

It is fitting that IBP should include a section aimed at the worldwide comparative study of human adaptability. IBP as a whole is concerned essentially with the functional relationship of living things to their environments – those living things still existing in relatively natural habitats and those in more disturbed or artificial conditions. An analogous approach can be made to the ecology of mankind. At this stage of human history vast changes are affecting the distribution, population density and ways of life of human communities all over the world. The enormous advances in technology make it certain that many communities, which have been changing slowly or not at all, will relatively soon be totally transformed. We are in a period when the biology of the human race is undergoing continuous change, measured in terms of health, fitness and genetic constitution.

IBP provides a great opportunity to take stock of human adaptability as it is manifested at the present time in a wide variety of terrains, climates and social groups, to deepen our knowledge of its biological basis and to apply this knowledge to problems of health and welfare. To do this satisfactorily for communities ranging from the very simple to the highly industrialised, requires an integrated approach and an application of methods drawn from many fields, particularly those of human environmental physiology, population genetics and developmental biology, aided by auxiliary disciplines, for example in medicine, anthropology, ecology and demography. Further discussion of the scope of the HA programme will be found in the book edited by Paul T. Baker & J. S. Weiner (1966).

Categories of research activities

The problems of human biology which are appropriate for study within IBP are manifold. National contributions should be selected from amongst the following categories in accordance with the interest and resources of the countries concerned.

Survey of sample populations in conformity with a world scheme. The general aim is to carry out as rapidly as possible surveys on a wide geographical range using standardised methods. Surveys are intended to make good deficiencies in our present knowledge of the distribution of important population characteristics. A knowledge of the distribution of these characteristics will, in itself, throw light upon many problems of human variability, adaptation and welfare. Extensive studies are required to determine gene frequencies of known polymorphic systems on growth and physique.

Intensive multidisciplinary regional studies based on habitat contrasts. The general aim is to elucidate physiological and genetic processes concerned in adaptation and selection in relation to climatic and other environmental factors. The multidisciplinary approach to a particular research problem and area would necessarily be based on an integration of the following components:

(*a*) basic socio-demographic assessment of the community for sampling, genetic and other purposes;
(*b*) basic assessment of the environment;
(*c*) general survey of genetic constitution;
(*d*) assessment of medical status of the subjects;
(*e*) assessment of dental condition;
(*f*) assessment of nutritional condition;
(*g*) background description of the daily and seasonal activities;
(*h*) assessment of physique and growth;
(*i*) assessment of working capacity as an index of fitness;
(*j*) environmental physiological studies;
(*k*) additional genetic studies.

Every population chosen is to be studied if possible by this comprehensive multidisciplinary approach. The particular problems of interest will determine which elements are to receive special and sustained attention. The studies may be grouped for convenience as studies with emphasis on environmental physiology, high altitudes, genetic constitution, nutrition, growth and physique, fitness (working capacity and respiratory function). Clearly certain of the special investigations proposed below could find a place in these multidisciplinary studies.

Special investigations on selected populations. A number of problems requiring study but on a less comprehensive basis than that of intensive studies described above have been selected for the consideration of national committees:

43

The Evolution of IBP

(*a*) studies of physiological fitness (working capacity and pulmonary function) of particular population samples. Three groups in particular are of interest: urban industrialised populations, non-industrialised populations, and athletes.
(*b*) disease as a selective agent of genetic constitution;
(*c*) particular socio-demographic factors affecting genetic constitution;
(*d*) other factors controlling population dynamics;
(*e*) special nutritional studies based on habitat and dietary contrasts, including the effects of the introduction of novel foodstuffs.

Ancillary medical and epidemiological topics. It is feasible in many cases to include in the above categories certain observations which are complimentary to current WHO interests, for example:

(*a*) surveys of blood pressure in relation to age, sex and occupation;
(*b*) haematological data (total red cell counts, etc.);
(*c*) antibody levels in blood;
(*d*) certain blood constituents (e.g. phospholipids, plasma proteins);
(*e*) congenital defects using a standard check list.

Phasing the programme

Phase I will include the assessment of methodology, leading to production of a handbook of agreed methods, research required for establishing the methodology, a training programme, formulation of the definitive programmes for Phase II, and also pilot, design and feasibility studies. During Phase II the survey work and multidisciplinary regional studies, as laid down under the above four categories of research, will be carried out by national or international teams in accordance with their national programmes. The categories are chosen to take account of (*a*) the need for urgent international action for the rapid accumulation of certain biological data; (*b*) the need for a multidisciplinary co-ordinated and sustained approach in certain 'key' habitats for the solution of basic problems; (*c*) the great variation in scientific personnel and other resources available in different countries; and (*d*) the relevance of the programme to health and well being.

The categories have been devised so that the research activities can be pursued at two levels – as 'extensive' surveys, by participation in worldwide co-ordinated studies, and as 'intensive' long-term regional surveys requiring internationally standardised methods.

44

Every endeavour has been made to ensure that HA is flexible in that free choice is given to the different countries to choose topics from a wide range, and that there is uniformity and compatability in the proposals in that similar topics will be investigated in different places using standardised methods. In this way systematic information will be obtained on many aspects of adaptability which can be related, over a wide range, to geographic and climatic conditions, nutrition and disease and other ecological factors. At the same time, new opportunities should be opened up for the study by many laboratories of the mechanisms underlying the processes of acclimatisation, physical fitness and genetic selection. It is intended also that many of the investigations will be extended to studies bearing closely on problems of health and welfare.

Use and Management of Resources (UM)

In the broadest sense, the wide scope of agriculture, fisheries, health and harvesting of domesticated and wild biological resources, are included in this section, but a much more limited approach needs to be made if reasonable achievements are to be realised. Wherever possible the work of such international agencies as FAO and WHO should be complemented in the fullest possible co-operation. With this in mind areas are proposed where international effort along basic biological lines might be expected to achieve success within the time limits of IBP. Nevertheless, the main themes include diverse activities requiring the services of numerous disciplines.

Plant gene pools

The widespread interest in the genetic diversity of plants of actual or potential use goes back to Vavilov's discoveries of geographical centres of genetic diversity. The interest was focussed then, as it is now, firstly, on plant introduction and plant breeding; secondly, on evolution and taxonomy. Since then collections have been established in various countries. Depending on the emphasis given to practical or to evolutionary ends, they were obtained prevailingly from plant breeding and experiment stations or through plant exploration. Such collections contain some or all of the following material: advanced cultivars, primitive cultivars, cultivars with special connotations in genetics, physiology or pathology, wild relatives of domesticated species, wild or 'semi-domesticated' species of actual or potential usefulness in pasture, forestry, soil conservation, etc.

Several botanical gardens and the Commonwealth Agricultural Bureau took an early interest in promoting exploration and seed exchanges. More recently, FAO has stimulated international collaboration and the Rockefeller Foundation has assembled and distributed to many nations large collections of certain cultivated plants. Such activities continue to serve a useful purpose in extending the use of genotypes of direct economic significance. It is, however, clear that other important aspects, and in particular a stocktaking of primitive and wild material, its collection and preservation in the face of rapidly spreading intensification of land use, call for urgent and extensive co-operation.

It is now realised that the usefulness of the collections is restricted by ignorance of the nature of plant adaptation, involving ecology, genetics, and micrometeorology. This problem calls for the collaboration of many institutions in diverse natural environments and for help from controlled-environment laboratories. It also calls for collaboration from Sections PT, PP and CT; and, equally, could not be attacked successfully without the closest collaboration with FAO.

IBP's work is planned under two headings, as follows:

Exploration, assembly and conservation of genetic stocks. Varieties produced by plant breeders can be excluded because they are represented in many institutes, they are relatively ephemeral, and catalogues, descriptions and exchanges are the concern of large national institutions and of FAO. It is proposed therefore that IBP activities be centred upon primitive races and wild relatives of cultivated species. Such material is of great scientific interest; it is a reservoir of genes and gene combinations of potential value in plant breeding; it has not been widely and systematically surveyed and collected on a world scale; and some of the wild relatives of cultivated species are regarded as weeds and so are threatened by the advance of civilisation.

The action programme should consist of the following: publication of a methodology handbook for the exploration, utilisation and conservation of plant gene pools; the stocktaking of relevant existing collections, in association with FAO's computerised record system for generic stocks; collecting expeditions in key regions composed of scientists from the region and specialists from elsewhere; the encouragement of live collections associated with active research; preservation of material in the form of seeds where this is possible, because this greatly reduces the risk of changes or losses though biological or mechanical contamination, mutation and parasitism; conservation of genetic variation *in situ*, which presents a problem akin to the conservation of habitats and communities which is the function of Section CT; 'type areas' would have to be numerous and widely dispersed.

Evaluation of plant resources: biology of adaptation. A plant collection is used by applied geneticists in two separate though interacting ways. Firstly, it is a reservoir of characters which are distinctive, can be readily observed, and are more often than not simply inherited. Secondly, the genotypes can be used to improve the adaptation and therefore the productivity of the plants. Such adaptations are usually based on many genes and therefore show complex heredity.

The first of these requires no support from IBP, but the second poses the question: in what manner can an understanding of adaptation assist in the transfer of potentially valuable genotypes between environments, either directly or by combination of gene blocks through hybridisation? This question raises problems of classifying genotypes in terms of productivity in the widest sense; environments in terms of physical parameters meaningful to plant development; and interactions of genotypes and environments.

The following crop plants are selected as suitable for wide study: hexaploid wheat, chickpea (*Cicer arietinum*) and rice. In the plan of investigation, there will be two distinct stages. Stage A will consist of field experiments which will aim to ascertain, for each of the three selected crop plants, the range of adaptability and the specific varietal adaptations in a widely representative genetic pool. It should then be possible to select a well defined series of, say, ten varieties representing different grades of adaptability and types of adaption. These will be used in Stage B for studies of the physiological, genetic and biochemical basis of the adaptability pattern revealed in the field. Where feasible the experiments in Stage A should include winter–spring, long–short day, early–late flowering or maturing types. A first series of experiments should be restricted to well known and identified cultivars of present-day cultivation. This would exclude primitive land races and populations, as well as more specialised genotypes. The fascinating problem of the relative adaptability of primitive and wild types could well be approached in a separate series once the more general experiments are under way.

Biological control – the control of biota by other biota

This research programme should be directed towards the elucidation of basic problems connected with selected topics and with biological control generally. Each topic should be regarded essentially as a co-operative research enterprise, the main lines and methods of which are planned in advance. To this end the programme should allow for the possibility of research being conducted in both temperate and tropical areas and in all the main geographical regions of the world. It will be advantageous if some of the topics selected can be fitted into existing national programmes. The programme should be restricted to a few pests in order to concentrate the effort and the depth to which the research is pursued. At least one of the pests should be important in relation to a major food source in developing countries.

With the above principles in mind, the following major pests are proposed for study: aphids of world importance, e.g. *Myzus persicae* and *Brevicoryne brassicae;* lepidopterans that attack the fruits and stems of apple and pear trees, e.g. Codling moth and Tortricidae; fruit flies of major international economic importance; rice pests, especially stem borers; and spider-mites of the *Tetranychus telarius* complex. Scale insects of the family Diaspididae, which are widespread throughout the world and of great economic importance, will also be considered.

For each of these it is planned to prepare a comprehensive review of present knowledge, to prepare a methodology manual with details of procedures for the study, and to convene working groups to finalise each methodology manual and to make practical work plans. A proposal from the USA to establish an international biological control training centre is strongly supported. This theme also includes plans for the preparation of a handbook on the general ecology of biological control, and for the encouragement of support services, particularly in taxonomy and in computer processing of data.

The development of biological resources and nutrition

Section UM together with Section HA is concerned with human nutrition, and, in the light of recent surveys, will give particular attention to foods that are rich in protein. This will involve work on methods for increasing the supply of conventional protein foods, improving the quality or digestibility of the protein in conventional foods, and the development of methods for making novel edible proteins. Examples of proposed studies are as follows:

Research is needed on the protein content and quality of used and at present unused plants. Concerning microbiological resources a proposal made in association with ICRO to establish mobile microbiological field-stations, operating in co-operation with technically advanced laboratories would offer a valuable approach.

Protein of high nutritional value can be satisfactorily extracted from many species of leaf and the technique should be of value in many protein-deficient regions, particularly in the wet tropics.

Strains of wheat and maize with improved amino-acid composition are already known. Differences in the amino-acid composition of ground-nut strains will be sought. Other projects will include the production of unicellular algae, culture of micro-organisms on hydrocarbons and vegetable wastes, the separation of protein from oil and fibre in the coconut.

There are also proposals for research on 'traditional' methods of food preservation often involving fermentation, designed to make them more trustworthy and effective.

Novel foods for farm animals will also be studied. For example, by-products of industrial operations and agricultural waste provide possible sources. Search will be made for improved strains of protein-rich seeds and tubers, of which several are limited in their utilisation because of the poor quality of the protein or the presence of detrimental compounds. Studies are planned on deficiencies and excesses of nutrients in animals in relation to both short- and long-term effects and with particular reference to mineral nutrients. It appears that the principal limiting factors in large areas of the tropics, especially in the lateritic soils, are those associated with imbalances of mineral nutrients, e.g. F, Zn, Mn, Fe, in plant products.

The principal statements of nutritional requirements of domestic animals have been developed in the temperate climates and differ between countries. The applicability of such statements under widely differing environmental conditions has been questioned and it is proposed to conduct a study of animal performance using the nutrient requirement statements as guides.

Differences between wild and domesticated large herbivores, have been shown in physiology, biochemistry and microbiology of digestion, and nutritional requirements, efficiency of feed conversion, nitrogen and water economy. Naturally occuring contaminations (e.g. myco-toxins) are factors influencing nutritional performance of animals, as are also pesticide residues in feedstuffs. Research on such questions now needs greater effort to secure co-ordinated progress.

It is a well known practice of agriculture to increase the efficiency of feed utilisation in grazing by following one species of animal, such as cattle, by another, such as swine. The examination of possible extension of such practices, including the involvement of micro-organisms and fish in such food chains, is a matter of concern and is included as part of the programme.

The IUNS will assist IBP in the development and carrying out of the nutritional aspects of the programme, both on human and animal nutrition.

3. Preparations

Period of IBP

When the possibility of a biological programme was first considered by the Executive Board of ICSU in October 1959, it was referred to as an International Biological Year, following the precedent of IGY; but the Executive Committee of IUBS discussed the proposed IBY a month later, and recognised that an operation lasting only a year, however intense the work involved, would be too restricted in time for the collection of a significant body of biological information. With seasonal changes, and variations from one year to the next, the collection of biological data often needs to be repeated at appropriate time intervals, so that a three-year operational term, preceded by some years of preparation, was proposed at that time by IUBS.

By the time of the inauguration of IBP in July 1964, there had already been several years of preliminary planning, as related in Chapter 1, but it was agreed that there must be three more years to be labelled 'Phase I – Preparations', and then 'Phase II – Operations' should last another five years, making eight years in all. Later, when the programme was well under way, it became clear that the volume of results was such that a further period of digestion and publication would be essential. So there were added two more years as 'Phase III – Synthesis and Transfer', thus turning the whole programme into a decade. Thus the period of IBP coincided with that of the International Hydrological Decade (IHD), 1964–74.

Now that decade has ended, but it cannot be claimed that the job is yet complete. Just as the IGY, which started as a year and turned into more than two years (July 1957–December 1959), had to continue its work subsequently in order to ensure the publications and use of the IGY data, so arrangements had to be made for the effective conclusion of IBP, especially in its international publications. As from July 1974, when SCIBP was dissolved, an IBP Publications Committee was set up by ICSU and will operate until the end of 1976 with a small office at the Linnean Society of London. By then it is expected that the main tasks entailed in winding up IBP will be completed.

The pro- and anti-IBP camps

As soon as the substance of the programme became known and widely distributed, environmental biologists around the world tended to divide into two camps. There was the pro-IBP camp, which looked at the prospects with some excitement. At last there would be opportunity, they thought, on a co-operative and co-ordinated basis, to examine some of the biggest problems facing the world – the future of man in his environment. Moreover, there seemed to be a chance of getting finance for some of these big issues for which, heretofore, finance had been given with reluctance either nationally or internationally – issues like the deterioration of land through inappropriate use, the overuse locally of fertilisers and toxic chemicals, eutrophication of inland waters, the effect of dumping waste in the sea. On the human side great issues had already loomed up, like the population explosion and genetic differences between races, on which it was difficult to organise research on an adequate basis because of the political overtones.

At that time also ideas were beginning to circulate about the application of systems analysis in ecology, applying mathematical and other techniques which had been developed from the relatively precise situations met with in the physical sciences and engineering, to the far more complex and less easily quantified ecosystems of the biosphere. To tackle this, massive data were required on biological and environmental variables, and this could only be done by research teams of considerable size, well organised and well directed; it was impossible under the old individualistic approach on which the science of ecology had been built. There was a general feeling too that, whereas cell biology and molecular biology had made great strides since the Second World War, and were providing new fundamental knowledge on which to build the applied sciences, the same could not be said for environmental biology. The practice of medicine and the breeding of plants and animals had benefitted from fundamental biological knowledge in somewhat the same way as engineering and technology had benefited from physical and chemical research earlier in the century; but the fundamental basis for land and water use for the purposes of agriculture, forestry and fisheries was not so well secured. IBP, while focussing on biological productivity, could perhaps start a new movement in environmental biology which would have important applications in the future.

Then there was the conservation lobby among biologists who recognised that national efforts were in total nowhere near adequate to meet the world's future needs for national parks, and nature reserves and 'field laboratories'. Biological 'bench marks' in the form of species, habitats and whole ecosystems were being lost at an alarming and accelerating rate, corresponding with increasing population and use of natural resources. Biological conservation needed a better and more fundamental scientific basis, which could be provided only at the international level.

The anti-IBP camp, on the other hand, consisted for the most part of biologists who were quite happy under their national scientific regimes. They were getting on well with their science and did not welcome new ideas being bandied about by international groups which tended to suggest new lines in which their work should develop. They saw danger moreover that new funds, as they became available, might be deflected into new IBP activities rather than be added to their own institutes. The anti-IBP camp tended to consist of what had become the Establishment in biological research, well heeled, and for the first few years of the programme's history it was considerably larger than the pro-camp. Between the two, of course, there was the indifferent centre consisting of scientists who cared little about what was going on outside their own environment, and tended to class IBP somewhat disparagingly as another bit of 'bio-politics', from which they would be happy to get support, if any support was on the go.

Although difficult to assess in any detail, there is no doubt that during the early years of IBP there was a considerable geographic bias in the distribution of these three groups of biologists. Most of the pros, at least of those who tended to shoulder the initial burdens, were from Western and Eastern Europe. Czechoslovakia and Poland were particularly active from the very beginning and were successful in gaining significant official as well as academic support. The United Kingdom, once the early indifference referred to by Waddington on page 8 was overcome, took a particularly prominent place through interest from the Royal Society of London. Indeed, during those early years one sometimes heard the jibe that IBP stood for 'International British Programme'. Japan was also early in the field and thorough in its actions, and so were the Scandinavian countries. It is probably easier for small countries to move into a thing like IBP than for the very big ones, but in due course the giants, USA and USSR, came fully into the picture and made the largest contributions, whether assessed in terms of manpower, publications, or finance.

As far as the developing countries are concerned, Phase I of the IBP took place during the years when the 'Wind of Change' was blowing vigorously through Africa and other former colonial regions. There was much reorganisation going on in colonial universities as well as in government departments and scientific institutes. Whereas the planning phase of IBP was picked up by enthusiasts in a number of such countries, it was difficult to sustain enthusiasm later in the face of many other demands for the strictly limited funds for science. This is not to say, however, that there were not a good many successful IBP projects in developing countries, as will be apparent from later chapters and the subsequent volumes in this series.

Perhaps the international organisers of the programme should have gone out in the early stages to publicise what they were doing and what they were asking other biologists to do. It was hoped that IBP would soon advertise itself from achievements, and in any case a publicity campaign was impossible since the international side of the programme has been run throughout with minimal finance and only rarely and for short periods has it had any professional assistance in public relations. The discarding of the original seventh section of IBP, entitled 'Public Relations and Training', was perhaps a mistake. Anyhow, outside a limited circle, knowledge about what IBP was and what it did has been limited up to the end, and this goes for scientists as well as the general public.

With hindsight it can now be appreciated that Phase I of IBP served a good purpose, not only internationally in working out the details and methods for the operational phase, and nationally in the preparation of plans and projects, but also in the maturation of the very idea of an IBP. During the course of those three years from 1964 to 1967 many biologists from the anti-IBP camp joined the pro camp and a few of the pros joined the antis. The indifferent centre drifted either way, but mostly towards the pros, while in some countries some of the stronger elements of the anti group became ardent leaders of the pros.

National activities

For a time it was hoped, and in some quarters almost assumed, that there would be some kind of international research fund which would help to finance core projects. However, it became apparent at an early stage that any money at the disposal of SCIBP was unlikely to go further than to support part of a modest organisation for international co-ordination. All actual research would have to be organised and financed nationally; and as national committees got to work and to plan their own contributions, new divergencies in approach became apparent.

A bit of competition in the preparation and publication of national programmes was encouraged, but to begin with there was a feeling of

54

despondency in some of the more developed countries. There was so much scattered research already in progress on biological productivity and human adaptability that there did not seem to be room for much more. Thus in some countries there was a tendency to give lip service to the ideas of IBP by listing in national programmes those pre-existing and on-going research projects in universities, government and other institutes, that appeared fit with the international published programme. Some other countries took the opposite view and regarded IBP as an opportunity to break away from the pre-existing systems and to include in their national programmes nothing but new projects, newly staffed, equipped and financed. Ultimately most national programmes became a blend of these two approaches: the criterion for inclusion of a project, both nationally and internationally, tended to become whether its emphasis was on advancing knowledge of fundamental ecological principles rather than the development or application of principles which were already known.

In all the drafting and re-drafting of plans and projects, the people with the real power over what research was to be undertaken, namely the national funding agencies, naturally took a part. The decade of IBP, and particularly this preparatory phase, coincided in many countries with a close scrutiny of the place which scientific research should have in the total national economy. Changes in funding systems were being introduced and it was inevitable that a new programme asking for finance, which had emerged from the non-governmental sector of science, took a rather low place in the priority stakes. The result in some countries was unfortunate delay in funding IBP projects, and this sometimes led to loss of interest among their promoters. On the other hand some advanced countries were quick to appreciate the advantages which might accrue from this initiative in connection with the whole problem of the human environment, which was then beginning to attract interest. In such countries a modicum of funds was voted, usually to be administered through the national academy of science, and where this happened the national IBP committee could really get to work.

However, the experience of IBP brought out clearly the fact that, while in some countries national organisation for financial support of research provided a framework for the support of international programmes, in other countries it did not. The effectiveness of procedures in this respect did not seem to be directly related to economic strength or political system. Where there was no adequate framework the difficulties were not limited to making appropriate contributions to international co-ordination, but they influenced the funding of national research as part of an international programme, and limited effective communication in the absence of travel funds. Since international science programmes are likely to increase, and to be of benefit to their participants, the need to provide recognised channels for their support, where they are inadequate, needs emphasis (see, for example, Frankel, 1972).

International organisation

Here we are concerned with the international aspects of Phase I, and in this it is impossible to exaggerate what IBP owes to the sectional leaders, the international conveners and theme organisers, none of whom have received any financial recompense, and many of whom have carried through their work with great efficiency but with the minimum paid scientific or secretarial assistance. Some sections have adhered to the same international conveners throughout, namely PP, CT, HA, and UM. Section PT had a major change when François Bourlière took over the Presidency of SCIBP from Jean Baer. The aquatic sections, PM and PF, adopted a rather different line and handed on the prime responsibility of convenership from time to time. Section UM, being less cohesive in its subject matter than the others, delegated much of the responsibility to the international theme leaders.

The principle of maximum delegation with minimum direction from the centre, provided the job was carried out effectively, has been a guiding principle followed throughout by SCIBP and its Central Office. It was recognised that the sectional arrangement allowed full rein for the differences in scientific discipline which are associated with the seven subjects. Attempts to bring them closely together might retard rather than encourage progress. On the whole this policy was successful but it also had its disadvantages, for the division into sections, and the division of each section into themes, each consisting of a group of national projects numbering from about ten to thirty, has tended to encourage a separatist approach. This has not invariably been the case but, as subsequent chapters will clarify, the failure of IBP to bring all its component subjects together into a total ecology – the ecology of human beings as well as the ecology of animals, plants and their environment – is one of the lessons which the programme has to teach.

Once the international section conveners, several of whom had taken leading parts in the pre-IBP discussions and planning, had been appointed by SCIBP, it was left largely to them to select their international section committees, representative as far as possible of subject specialities and geographical regions. These committees were particularly active in Phase I; during Phase II it became less necessary for them to meet although they continued as a major element in the communication system; in Phase III they were dissolved in favour of editorial working groups arranged around the subjects of each volume in this series. The list of conveners, scientific co-ordinators and of section committee members who have served IBP so well, is given in Appendix 1, together with the officers and members of SCIBP.

The international pattern of section committees responsible to a main IBP committee was followed also by many of the larger participating countries, although some found it convenient to amalgamate several sections. Thus those sections concerned with terrestrial subjects – PT, PP and CT – were sometimes grouped in one national sub-committee, as were PF and PM dealing with the aquatic environment. On the other hand HA, consisting of biologists who have been drawn largely from schools of anthropology and physiology, was kept separate.

Achievement and failings of Phase I

With this background we can now consider what was actually achieved during the three initial years of IBP that constituted Phase I, and where it failed. Like all other aspects of IBP this must be considered in two parts, the international part consisting of activities for which SCIBP was responsible, and the national part for which the national committees were responsible.

On the international side, there were two main objectives: the first was to work out the programme in detail, section by section and theme by theme, not forgetting that some subjects were bound to overlap sectional and theme boundaries; and the second was to make available to all participants in the programme and to any other scientists interested, the best and most up to date methods of research, at different levels of sophistication. In most subjects these two objectives could be achieved by consulting the same groups of people, and this involved drawing together leading biologists from many parts of the world for a substantial number of consultations.

The technical meetings which were arranged (see Appendix 4) were only rarely symposia at which scientists read papers to each other; nor were they the executive kind of meeting designed to take decisions. They were something between the two, each of them convened for a specific purpose connected with preparing the programme. Often this was to lay the groundwork of a handbook on methodology: indeed, some of the most successful of these meetings had draft chapters on methods as working papers. These drafts were discussed before specialist audiences, and were subsequently revised and brought together under the guidance of selected editors. The series of handbooks which resulted (see Appendix 5a) is largely devoted to methodology in the IBP subjects and has of itself become a considerable contribution to international collaboration. In some cases studies of methods of research, which resulted from these early meetings, were published through agencies outside the IBP hand-book series, a notable example being the large and comprehensive photosynthesis handbook edited by Sesták, Cătský & Jarvis (1971), which is listed in Appendix 5b.

Although words like 'standardisation' and 'conformity' of methods have sometimes crept into these handbooks, the point must be made that there was never any intention on the part of IBP to *standardise* methods, for this would retard the progressive evolution which most of the subjects concerned are currently undergoing. The object was to *recommend* methods which were likely to give intercomparable results, wherever in the world they were used. The fact that several of these handbooks have, during the course of IBP, needed extensive revision, and have been reissued as second or third editions, is a measure of their usefulness. In one or two cases the ideas which were prevalent in the early years of IBP and were put together into handbooks have been improved and replaced during the course of the programme to such a degree that a few of the early handbooks are being replaced by entirely new volumes.

A number of the technical meetings held during Phase I, in addition to achieving their specific purposes, provided a forum for the presentation and discussion of new ideas which were deemed worthy of publication in their own right. This non-series of volumes, published in a variety of formats and in several different languages, has grown to fill a substantial shelf.

While these activities were proceeding at the international level, the number of participating countries was steadily growing. In the process of preparing and organising their own IBP projects, some of the larger countries held technical meetings and symposia under their own national auspices. Often foreign specialists from other IBP groups were invited so that not infrequently these national discussions took on international significance. Different countries interpreted their needs and potentialities

58

in relation to IBP in different ways, so there is no rigid uniformity in the selection and description of projects. This perhaps was as well for the entire subject matter of the programme was by no means uniform, and the research demanded many different approaches.

All projects in national programmes were considered by the appropriate international section committee or working group before their submission to SCIBP, but here again, it cannot be claimed that there was uniformity in the screening process. At this early stage it was usual in most branches to accept all projects which had relevance to the published description of the programme, provided there was reason to believe the research would be of high quality. Later on, as the programme became more focussed, some projects that appeared in the lists were found to have only marginal relevance.

In some cases two or more countries joined forces to produce joint projects, some of which proved to be highly successful. This indeed was one of the means of invoking the potentialities and enthusiasm of some developing countries in IBP: there are good examples in Latin America, Africa and the Far East, in which a developing country contributed the all important problem, the site, together with local facilities and junior staff, while a developed country contributed additional finance and scientific expertise. As a result such projects included a considerable element of scientific training and education.

During Phase I certain countries, having common problems and sometimes common frontiers, came together to form IBP federations. The most advanced of these was Scandinavia, where the chairmen of national committees in Denmark, Finland, Norway and Sweden came to meet regularly and arrange a division of labour in their national programmes, which were published in one volume. A parallel case, though for various reasons – political and financial – it never reached full fruition, was East Africa, comprising the countries of Kenya, Uganda and Tanzania. In the New World common interests were developed at several pan-American IBP meetings, joint projects were organised, and a newsletter, *Inter-American News*, was published. Furthermore, some pre-existing regional organisations proved very helpful, notably the Pacific Science Association. Another good example of regional collaboration was between countries which have particular interests in the Arctic, namely USA, Canada, France, the Scandinavian countries and the USSR.

A good many countries which were well advanced in their planning did not wait until the end of Phase I before initiating operations. Certain projects were operative by 1966, and in consequence a trickle of scientific papers resulting from IBP research began to show itself even before the end of Phase I, and was destined quite soon to become a considerable stream.

It soon became clear that field studies stimulated by IBP were bound to result in a large quantity of data for which, quite apart from their collection, the problems of storage, retrieval, and use through various forms of processing with the aid of computers and other equipment, would become acute. Thus the establishment of world data centres, a concept somewhat new to biology though well recognised in the physical sciences, became a talking point as part of the preparations. With one or two notable exceptions, particularly in relation to the conservation of habitats, the difficulties of deciding just what data should be provided for, and how data centres should be organised and financed, proved more or less insuperable, so the lack of international biological data centres at the end of IBP must be regarded as a disappointment. This is not to say, however, that the data did not become available or that the problem was not provided for in other ways. Indeed, vast quantities of data have been used in the primary publication of IBP results and many have been brought together in this series of international synthesis volumes. They will be available for reference in numerous national centres. But the establishment of major international data centres for environmental biology remains a problem for the future: it is satisfying that this problem is now being examined by CODATA which is taking IBP's experience into account.

The IBP process

Thus by the end of Phase I in July 1967, what may be described as the IBP process was well established. Broadly speaking, this consisted of first defining a problem, then bringing together a small number of the most competent specialists on the problem irrespective of nationality, and with their assistance establishing an action plan; thereafter that plan was put into operation, usually with a minimum of new finance, and finally the same or a similar group of specialists was called together in order to assess the results and fit them into the pattern of IBP as a whole. Through this process there was already growing by the end of the preparatory phase a coherent but flexible network of biologists covering a large number of countries, and they were communicating with each other about different aspects of the programme.

Following the practice of organisations in the ICSU family, the maximum amount of the work was done on an honorary basis by scientists who gave generously of their time and effort. It was recognised from the beginning, however, that the very breadth of the programme would necessitate whole-time international attention by a Central Office and section offices, and by the end of Phase I the Central Office was staffed by two scientists, and each of the sections except one had a whole- or part-time scientific co-ordinator, together with secretarial assistance. Meanwhile, a number of the larger participating countries had established their own IBP offices, related to their national academies of science, and had staffed them as appropriate. The scene for the operational Phase II had been set.

4. Operations

By the end of Phase I in July 1967 the seven sections of the programme divided into some eighty themes were active at the international level, and a substantial number of scientific leaders had emerged through the working groups, committees and technical meetings. However, the time it took for the ideas of IBP to get around the world, through the universities, institutes and individuals who were best able to make contributions, and to the national academies and national agencies which were capable of producing funds and organisational structure, had been underestimated. Owing to delays in communication and the need for back-checking and printing, it was not until 1969 that the Central Office was able to publish the comprehensive list of projects, section by section, in *IBP News*, nos. 13–20.

While the patterns of IBP research had emerged earlier, they did not come fully into focus until all the national projects could be tabulated. These patterns of research will appear theme by theme throughout this series of volumes. The object here is to paint on a broad canvas in a somewhat impressionistic manner, to indicate what happened in the programme as a whole. In doing this we will illustrate by examples, where possible in the words of some of IBP's scientific leaders. It was hoped to highlight in this chapter the achievements of the programme, but the contributors were understandably reluctant to do this until the results of each main part of IBP had been assessed in the volumes which are currently in preparation.

Although dividing lines are indistinct it is convenient to distinguish three main ways of designing ecological research and viewing the data which result therefrom: the first is through autecology, in which a single species or a small group of species is studied to determine its status in the whole environment; the second is through synecology, in which communities of species, plant or animal, or both, are studied in their relationships one to another; and the third is through the ecosystem approach, where the total biotic community plus its environment is the unit. Autecology can be undertaken by a single research worker or a small group. Synecology involves more workers, or at least the availability of specialists for reference concerning the different species which make up the community. Study of a total ecosystem involves not only the biological disciplines, but also the physics and chemistry of the environment, including rock, soil, water and atmosphere.

It is this ecosystem approach which distinguishes much of IBP research from what had dominated ecology before. Essentially it consists of the careful selection of a number of variables – biological, chemical and physical – about which data are collected, quantitatively as well as qualitatively. Thereby the ecosystem can be analysed in order to ascertain which factors and processes are important in causing the dynamics of the whole. In this, the application of systems analysis to biological systems has been one of the major innovations developed during IBP. Some would go further and say that it has been one of IBP's major achievements.

During the numerous discussions about systems analysis and mathematical modelling that were held at national and international levels during the progress of IBP, there was often an element of scepticism. Modelling a total ecosystem is a more complex undertaking than, say, an engineering system and can be of a high-resolution or low-resolution type depending on the extent of the system to be defined. Systems analysis in ecosystem studies is very important in stimulating the production of hypotheses, sorting them out and selecting the most appropriate for future study. The real criticism of 'ecosystems systems analysis' lies in the over-enthusiastic claims of some of its adherents. There are still some leaders in ecology who, while recognising the value of a systems approach in cases where sufficiently accurate data are available, consider that in ecological research as a whole, the time for this is not yet ripe. However, there is no question that the systems approach was an important factor in many branches of IBP, even if few IBP projects started out with a definite orientation towards it. Apart from the US contribution to the PT programme, which was systems-oriented at the outset, many of them started with other approaches in mind and subsequently attempted to re-orientate themselves towards dynamic systems models. Thus the objectives which it was hoped could be achieved within IBP by the systems approach are considerably more limited than those which might have been achieved by the adoption of a systems strategy from the outset.

It must be recognised that only comparatively recently have ecosystems been analysed in this fashion and that time and experience are now needed in order to fine down and perhaps to economise in the techniques which have thus been tested. Looking to the future, there is no doubt that the systems approach has come to stay, but we must be careful not to define it too narrowly. What is actively being advocated by some enthusiasts is narrower than the total strategy which has become available through the adoption of modern quantitative techniques.

The value of systems analysis as a tool of ecology will be a recurrent theme in many of these synthesis volumes. It applies to all sections, but has been developed particularly in the Biome studies of PT. Therefore it is well to start a general account of the operational phase of the programme with a summary of that section's work, in the words of its International Convener, J. B. Cragg.

Terrestrial ecosystems

'Several metres of PT progress reports fill my bookshelves. To reduce these to a few pages of text without doing injustice to many studies, or oversimplifying what are complex investigations, is no easy matter. Thus, this summary begins with an acknowledgement to the hundreds of scientists who have contributed to the PT activities and a plea for their understanding if the products of their work appear to have gone unnoticed at this stage. They will have their opportunity to put the record straight in the biome and theme synthesis volumes.

In the early discussions which gave shape to the PT programme, considerable emphasis was laid on the need for what were termed *minimum* and *maximum programmes*. The former were to include the measurement of a selected number of parameters and would not necessarily require the use of sophisticated apparatus or procedures, or involve a high level of biological knowledge from participants. Its main purpose was to guarantee that the final results on productivity would contain basic information from all the major biomes and particularly those with sparse populations and with few centres for biological research. In addition, it was considered that the *minimum programme* would give opportunities for training scientists in those areas of the world where biological studies were still at an early stage. *The maximum programme* was to be limited to a number of places staffed by highly trained personnel with adequate technological support. As IBP got under way the majority of the national committees showed enthusiasm for highly complex biome and theme studies and the *minimum programme*, apart from a small number of important exceptions, was gradually forgotten.

Final agreement on the general pattern of the PT programme was reached at a meeting held in Poland in 1966 (see *IBP News*, no. 7).

Various projects were classified according to Biomes and Themes.*
In the biome category there was a preponderance of sites in the temperate
zone with woodland studies predominating, but with grasslands and
deserts taking their share. By 1970, some balance had been achieved by
the welding together of a number of Arctic and sub-Arctic sites to form
a Tundra programme. To these, at a later stage, were added sites with
somewhat similar habitat features but which lay outside of areas
normally defined as tundra.

Where possible, comparison was to be made between undisturbed
and disturbed areas. In practice, 'undisturbed' has usually meant 'least
disturbed' because few areas of the world, certainly those included
within the scope of IBP, have escaped from the effects of man's activities
either directly or via his domestic animals. In some cases comparisons
have been made but in others such comparative studies were not feasible.
In addition to investigations on disturbed areas, it had been proposed
that cultivars, appropriate to the climate and to the biome, should be
grown on or in the neighbourhood of natural areas to provide standards
against which to measure various features of organic production. Such
experiments have been tried in a limited number of places. For example,
within the Tundra Biome comparative investigations on cultivars have
been made at sites as distant from each other as Greenland and South
Georgia.

The individual national committees were responsible for the investigation of those sites lying within their jurisdiction, and considerable
differences in emphasis occurred between teams in different countries
and between the different biomes. Indeed, the comparison of national
reports and discussions at meetings of the PT committees and workshops
gave substance to a statement of Margalef (1968). Speaking of ecological
succession and the exploitation of ecosystems by man, he says: 'Ecosystems reflect the physical environment in which they have developed,
and ecologists reflect the properties of the ecosystems in which they have
grown and matured. All schools of ecology are strongly influenced by a
genius loci that goes back to the local landscape.' These differing viewpoints, described by René Dubos as 'the genius of the place' are a
summation of the physical, biological, social and historical forces
which have shaped localities and their people. They have provided a

* As explained elsewhere the totality of IBP is broken down into *sections* and within each
 section a number of *themes*, each of which is composed of relevant *research projects*.
 However, PT has used a further distinction into *biomes*, such as woodlands, grasslands,
 which included a full integration of many different approaches, and *themes* which are
 more limited in concept [Editor].

source of diversity in all Phase II workshops. But, in spite of these differences in attitude and approach, there has been a marked desire throughout IBP for the participants to seek comparability between field investigations. A degree of international co-operation between terrestrial ecologists has emerged which is best described by a quotation from a biome report: 'An important milestone in mutual respect and trust has been established which has led to sharing of data, collaborative comparative analysis, and future planning of environmental research programmes. This human resource is too valuable not to be utilized in future environmental programmes. Many years are required to establish such a functional network of scientists, but it can be destroyed in a very short time without continued nurture.'

Many of the sites selected for intensive studies during IBP were not selected because of their special suitability of IBP but because a body of scientific knowledge already existed for the site and sophisticated experimentation was possible. This accounts in part for the lack of adequate coverage of many biome habitats. In some national programmes, however, in Poland and the USA, for example, sites were actually selected to give the best possible spectrum of habitat types.

By the time that Phase II began, most teams of IBP workers were working with certain general aims in mind, even if an integrated overall plan had not been formulated in specific terms. An awareness of important gaps in knowledge such as, for example, the contribution of the ground flora of a woodland to total organic production, the consumption of plant materials by invertebrates, the interrelations of microflora and microfauna, the distribution of certain elements such as nitrogen and potassium, were some of the problems which helped to establish bridges between scientists with different special interests. It was at this stage in IBP that the 'big-science' concept in biological research gradually emerged. It became obvious that if the dynamics of even a relatively simple natural ecosystem were to be understood, then individual projects on IBP sites would have to be integrated in accordance with certain biological principles. A managerial approach, using systems dynamics techniques became increasingly necessary.

Workshops in all biomes began to emphasise the systems approach and PT has served as a major 'proving ground' for systems ecology. The aim is to provide an understanding of how a total ecosystem functions and, by changing the parameters of various components using systems dynamics techniques, to predict the responses of an ecosystem to natural or man-made stresses.

Just as the application of mathematical techniques to the study of animal populations has clarified many issues by providing a framework within which hypotheses can be formulated and tested, so the advent of the computer and the development of systems dynamics has provided the ecologist with fresh insights into the structural and functional relationships which exist within ecosystems. There can be little doubt that the results now being obtained from some of the biome studies will provide a basis for the development of reliable management procedures. At present the models which have been constructed are of a simplified type. This is not a criticism of the approach because the simple model is an essential first step. In fact, one of the most important features of the growth of systems analysis within IBP has been the diversity of approaches even within a single national programme.

For example, in the USA programme, which has led the field as far as systems applications are concerned, there are contrasting features in the way in which the systems approach has been applied in the different biomes. In the Grassland and Tundra Biomes, studies were organised around a general model to cover the major aspects of the ecosystems under investigation. Later, parts of these low-resolution models were studied as sub-models. Such a series of sub-models is shown in Fig. 3, which represents the Point Barrow wet coastal tundra ecosystem. The USA Aridland Biome, instead of starting with a general model, developed three sub-models dealing respectively with plants, animals, and soils and these were used to explore specific problems. The deciduous forest biome began with a series of sub-models, each dealing with fundamental processes in the forest ecosystem. Only towards the end of Phase II have major efforts been made to fit the sub-models together to provide a functional picture of the total ecosystem. Whilst it is unlikely that the present IBP synthesis will provide a comprehensive quantitative account of the major functional relationships in a given ecosystem, the systems studies will at least provide word-models with a degree of precision which will help to give a greater degree of predictability to management procedures. Model building is shown amusingly in Fig. 4.

The greatest effort in PT has gone into the study of ecosystems dynamics. The majority of theme projects as distinct from biome-oriented investigations proposed in 1966 failed to attract sufficient support and only four became fully operational. Each of them – decomposition and soil processes, small mammals, granivorous birds, social insects – were concerned with problems of considerable importance both for understanding major aspects of ecosystems dynamics and for

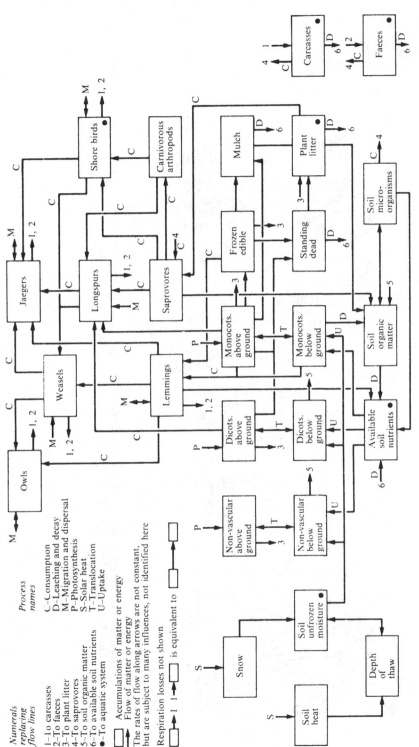

Fig. 3. Example of ecosystem analysis. Flow diagram of the tundra ecosystem near Point Barrow, Alaska.

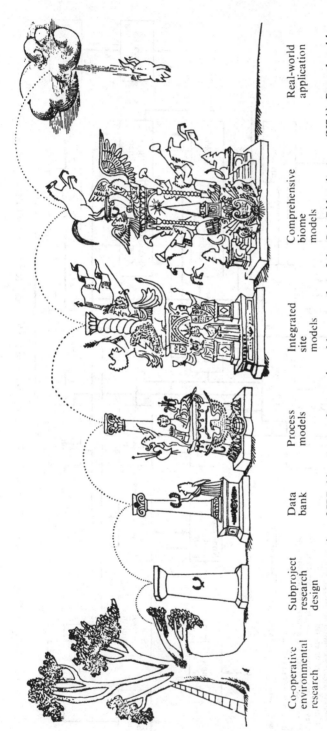

| Co-operative environmental research | Subproject research design | Data bank | Process models | Integrated site models | Comprehensive biome models | Real-world application |

Fig. 4. A tongue-in-cheek representation of IBP, biome research as envisaged by a member of the Oak Ridge site team (USA). Reproduced by permission from *Eastern Deciduous Forest Biome* US IBP Analysis of Ecosystems Newsletter no. 9, March 1972.

providing information urgently required either for pest control or for ecosystem management.

The decomposition and soil processes theme was concerned with phenomena common to all ecosystems. Input into the theme has depended on a close integration of studies on specific sites within the different biomes. The synthesis of results is being obtained by a team drawn from the various biomes. Measurements have been obtained on: input of organic matter and the effect of abiotic and biotic factors on its decomposition; changes in physical and chemical composition of materials during decomposition; the structure and function of soil and litter communities associated with decomposition; the interaction between different components of the microflora and fauna; and energy flow and the distribution and circulation of major chemical constituents. The systems approach is being utilised in the interpretation of the various data.

The small mammals theme has been the first component of PT to complete its synthesis. It has drawn together information from all biomes and represents a major achievement of IBP. For the first time, a fully integrated account of the structure and function of small-mammal communities on a world basis is available as a ground work for future studies and as an invaluable tool on which to base control procedures. Much of the work has been concerned with problems of population dynamics; physiological and behavioural aspects of species and groups; biological production and bioenergetics; the role of small mammals in chemical cycling; their special activities in different biomes; their control in natural and urban environments.

The social insects theme was originally intended to cover all major groups of social insects. Very soon after its inception its efforts were concentrated on two highly important and widely distributed groups, the ants and termites. The synthesis volume will bring together information on population dynamics, production, food and feeding activities, energy flow, their role in the functioning of different ecosystems and the bearing of these studies on the control of these organisms.

Much of the work of the granivorous birds theme has been concentrated on species of *Passer* which account for major depredations to grain crops by birds. Supporting studies are available for species of other genera such as *Agelaius*, *Quelea*, *Spiza*, *Corvus* and *Sturnus*. These comprehensive studies covering a range of ecosystem types from natural to urban situations, are now approaching completion. They will fill an important gap in our scientific knowledge of the activities of granivorous bird populations. Of major significance, they will provide, for the first time, a corpus of knowledge on which to base control procedures, many of which in their present form, the use of certain toxic chemicals for example, can create hazards for human populations. The evolution of these birds has been investigated and in particular, the mechanisms which have given rise to granivory and the emergence of certain species as 'animal weeds'. They have proved excellent subjects for the application of systems dynamics procedures, especially in the study of their population fluctuation, bioenergetics, and general role in ecosystems.

Looking at PT's achievements broadly, certain general comments can be made. The major outcome of the terrestrial biome studies is a considerable amount of information on the structure and functions of ecosystems which has been obtained on a worldwide basis by more or less standard procedures. Prior to IBP, the only available data on such matters as primary production, the biomass of different categories of consumers, the major pathways of energy flow, or the distribution of major chemical elements, were few and were not strictly comparable between many different sites. Now, as a result of IBP it will be possible to draw on information of a degree of reliability not previously available. Coupled with this considerable increase in knowledge has been a gradual evolution in technique, a change documented by the demand for revised editions of several IBP handbooks. The report on Quantities, units and symbols (Appendix 6) is a milestone (perhaps the authors of the report would prefer to call it a *borne kilométrique*) on the road to the full achievement of comparability in ecological studies.

It is evident that systems ecology, even in its present simplistic form, has considerable potential both as a tool for providing fresh insights into the functional aspects of ecosystems and as a method of improving management procedures. Some of the sub-models developed in all of the biomes have direct applications for the investigation of stress effects. Their usefulness in connection with the management of logging operations or in regulating grazing intensities has already been recognised.

One positive result of IBP which will have an important bearing on future environmental studies, is the emergence of a group of young scientists capable of looking at a complex problem in interdisciplinary terms. Whereas in the early stages it was difficult in many cases to achieve a team approach to an ecosystem study, the team concept has now been accepted as an essential procedure for certain types of ecological investigations. This does not mean that the role of the individual scientist has been demoted, for the degree to which a community effort is required depends very greatly on the nature of the problem.

Whilst IBP/PT has accomplished much, it should be regarded as no more than a feasibility study for a major investigation of the many processes which are fundamental to the maintenance of man's environment. The phrase 'feasibility study' is used advisedly because, even in those parts of the PT programme which have a five-year set of field observations, few of them contain more than two or three years of detailed study on named species, populations or habitats.

The failure to implement the *minimum programme* of PT on a world basis and to finance a larger number of *theme studies* is to be much regretted, and has affected the outcome in two ways. Instead of ensuring worldwide coverage, certain biomes and regions, and some important groups of organisms, are inadequately represented in the results. For example, the amount of work on large herbivores incorporated into the biome studies is very small. Furthermore, in concentrating on major and often highly specialised studies at a limited number of sites, readings for all the parameters originally listed for the *minimum programme* are not often available. SCIBP, in the final months of its existence, has realised the importance of *minimum programmes* and, acting on a proposal from the Co-ordinating Committee of the Woodlands Biome, has contracted with UNESCO for the preparation of a handbook on a Minimum Research Programme for Tropical Areas.

Another shortcoming of IBP concerns data banks. Most biomes made tentative plans for establishing centres where the often massive data collected could be stored, retrieved and processed on an international basis, and the grassland group in particular gave much time and thought to this. However, proposals which emerged could not be implemented for lack of funds and, as a consequence, the synthesis which will emerge from Phase III will fall short of what might have been obtained. Looking at the future it is of considerable importance to make available what are, for the most part, carefully collected data dealing with the structural and process components of sites scattered across the world. These data can be considered as IBP's 'renewable resource' for the research and management studies which are to follow. At present some data banks in fact exist: the Woodlands Biome, for example, has established two, one at Göttingen for inputs from sites in Europe, Africa and Asia, the other at Oak Ridge, Tennessee for Australia, North and South America. These and other arrangements are dependent on national as distinct from international funding so their long-term life cannot be guaranteed.

It is important, therefore, that the status of data banks, the types of data to be stored, their availability for research and other purposes, should be reviewed by an appropriate international organisation very soon.'

Process studies

Biological processes such as photosynthesis, the passage of water through plants, processes involved in various mineral cycles, consumption of plants by animals, have long been a central part of the science of ecology as well as of physiology. However, the international co-ordination of efforts in such subjects afforded by the Section PP of IBP has given a new dimension to their study. It has brought together not only scientists from different countries, but also specialists in fields which were previously relatively remote one from another, for example, terrestrial and aquatic biologists, microbiologists and plant taxonomists. There is now considerably less danger of excessive parallel research of the same kind arising through lack of communication among the scientists concerned. Certainly among the highlights of IBP achievements are the advances made in the two major processes, photosynthesis and nitrogen fixation by micro-organisms, on which the PP section focussed its attention.

The following account, in which some of this section's activities are brought into perspective has been contributed by Ivan Málek, the section's Convener, in collaboration with his two scientific co-ordinators, Jan Květ (photosynthesis) and Eva Hamatová (nitrogen fixation).

'Any scheme or model of an ecosystem or a biotic community is composed of compartments ("boxes") and transfer functions ("arrows"). These transfer functions may be defined in purely stochastic terms but it is more valuable to define them as time rates of the processes. Indeed, the structure and functioning of an ecosystem may be studied and understood as a result of a complex interrelationship of the various processes taking place in it. This is the essence of the "processes approach" to ecology, and the Section PP became, within the IBP structure, the main centre for pioneering such an approach.'

Physiological processes in the principal plant, animal and microbial populations constituting an ecosystem are crucial. With the emphasis on production ecology, two were selected as of major importance, namely plant photosynthetic production and microbial fixation of elementary nitrogen. At the beginning of IBP the idea was sometimes

74

expressed that for the study of the structure and efficiency of production in ecosystems existing knowledge of these two processes was, by and large, sufficient, but it was soon agreed that a self-contained co-ordinated study of these processes was necessary. Particularly important was a more accurate quantitative evaluation of their respective contributions to biological productivity under different conditions.

During IBP, knowledge of both processes and an apprehension of their respective roles in nature have been deepened substantially in the following ways:

(*a*) Individual problems were identified and international teams of specialists were concentrated on them in different ecological conditions.
(*b*) The methodology and technical approach to these studies was enlarged, deepened and elaborated.
(*c*) The respective roles which these processes play in individual ecosystems have been defined more exactly.

The PP section had a successful predecessor, the Commission for Ecophysiology in UNESCO, led by F. E. Eckardt, who organised two important symposia, at Montpellier in 1962 and Copenhagen in 1965, and thereby helped materially to launch PP as a separate section. The process studies became acknowledged as highly important in the biological productivity of water as well as land, so an intersectional working group for photosynthesis studies in different ecosystems, which brought together specialists from the PT, PF, PM and UM sections, achieved a logical integration with the rest of relevant IBP work.

From 1965 to 1967 progress was marked by meetings of the sectional committee or of its two sub-committees, referred to as PP-P and PP-N, in Prague, Copenhagen, Moscow, Orsay and Paris and Addis Ababa. PP-P became concentrated mainly on the ecology of photosynthesis in both crop and natural systems; work of PP-N remained confined to the process of nitrogen fixation by symbiotic or free-living micro-organisms. Some critics suggested that PP-N should also include the cycling of nitrogen after fixation, a suggestion which gained importance when the efficiency of protein production came into prominance as a world food problem. Nevertheless, the concentration of IBP's effort on primary nitrogen fixation proved wise since it absorbed nearly all the research capacity available. Development of these studies towards the wider problems of nitrogen cycling and efficient production of protein was however, pursued by the Scandinavian PP programme and also by UM's working group on novel protein sources.

Other proposals to extend the PP programme, for instance, an Australian proposal to study the processes involved in phosphorus cycling, were likewise resisted in order to avoid too great a diffusion of effort. From its early days, the PP section had a firm organisational backing in its secretariat in Prague, which was created by the Czechoslovak Academy of Sciences in 1965. It was divided into the two sub-sections in 1967, and remained in operation until the end of 1974.

Photosynthesis and solar energy conversion (PP-P)

When formulating the PP-P programme there was, as in PT, the question of the so-called "minimum" programme. In 1965, G. E. Blackman and the Deputy Convener, E. C. Wassink, proposed two kinds of simple experiments: seasonal changes in growth parameters of widely spaced young plants (sequential sowing experiments), and production studies on plants given various levels of daylight (shading experiments). There were also proposed experiments on the net photosynthetic production of crops sown at different densities, with the aim to find out what was described as "potential production" in a given region or site. In all these experiments, the plants required a lavish supply of water and mineral nutrients.

While such experiments were, in fact, conducted in a number of countries the main interest became focussed later on more sophisticated studies of photosynthesis of crops and other communities in relation to their structure, physiology and environment. Thus an Australian proposal, adopted in 1967, lists many kinds of studies relevant to this theme. Detailed measurements of photosynthetic activity under the influence of both internal and external factors form a framework at whose centre are the measurements of CO_2-exchange *in situ*, using both aerodynamic and assimilation-chamber methods. Biochemical and biophysical investigations of the photosynthetic apparatus are also included. This proposal had a wide following, and most of the research falling under the scope of PP-P came to be of that kind. The sub-section brought together representatives of various schools investigating plant photosynthetic production and greatly stimulated fruitful discussion and co-operation between them.

The focal points of such activity were technical meetings particularly that arranged at Třebon in 1969, which confronted the new modelling approach with the classical experimental approach, and that at Moscow (also in 1969) which brought together all the aspects of fundamental photosynthesis. The proceedings of these two major meetings, and also a handbook on methodology, and the new international journal *Photosynthetica* (see Chapter 6), although not published by IBP itself, remain a heritage of PP-P.

Later in the programme the work of an intersectional group on photosynthesis in different types of ecosystems was particularly valuable in bringing together the results of both IBP and non-IBP research in this field. It culminated during Phase III in the symposium at Aberystwyth in 1973 which laid foundations for the synthesis volume dealing with this topic and demonstrated how the "processes approach" can be brought into close context with the rest of ecology.

A survey of national programmes, made in 1968, revealed some 200 IBP projects on photosynthesis, but national participation in PP-P was rather varied. A few countries had very well integrated programmes. Some focussed primarily on the simple experiments, but most on more sophisticated work which was conducted on traditional crops, on novel ones like algal cultures, and also on wild-growing plant communities.

Much of the work was geared towards collecting data for mathematical models of photosynthetic production. Such models attempt to define the dependence of production and growth on both the environment and the internal status of the plants. They offer good prospects for predicting production under various sets of conditions and in various habitats, as well as in crop management for obtaining high yields. It is becoming important to compare photosynthetic production from agricultural crops with that found in both terrestrial and aquatic natural and seminatural communities, but such a comparison will be feasible only after the vast amount of IBP and other data has been evaluated.

It is difficult at this stage to define the achievements of PP-P, but three points of both theoretical and practical value may be mentioned:

1. The net photosynthetic production is perhaps less affected by photosynthetic activity as such than by other physiological processes determining the fate of the assimilate: respiration, translocation and distribution of assimilate and developmental status of the plant. In this context, both short-term functional and structural adaptations of the photosynthetic systems (from the chloroplast to the leaf canopy or algal suspension level) and long-term structural and developmental adaptations become very important. Contrary to the ideas of some systems ecologists, the species concept is highly important since the physiological response of the plants to various environments depends

markedly on genotype. The practical importance of such findings for agronomy and plant breeding is obvious.

2. A "non-IBP" discovery of the last decade – that two distinct types of photosynthetic metabolism operate in higher plants (C_3- and C_4-plants) – is of great importance for the interpretation of ecological measurements of photosynthetic activity. The two metabolic types differ in their response to solar energy supply, through the presence or absence of photo-respiration, as well as in the plant's anatomical structure. IBP data seem to indicate that C_4-plants occur more frequently in the warm climatic zones.

3. Comparisons between actual and potential photosynthetic production, the latter being often simulated by mathematical models based on climatic parameters, have revealed that plant communities only rarely operate at full rate, and none of them do so for the whole potential growing season. By removing the various limiting factors, such as water stress and mineral shortages, and by breeding more versatile varieties, the net photosynthetic production can be considerably increased. Maintenance of a favourable productive structure, intercepting as much as possible of the available solar energy, is perhaps the main factor in achieving yields. All agricultural and aquacultural practices ought to be evaluated from this viewpoint.

Biological nitrogen fixation (PP-N)

The results on the fixation of elementary nitrogen have to be evaluated from two standpoints: firstly, to what extent they have contributed to the deeper knowledge of the fixation process as such, and secondly, how far this deepened knowledge has enriched the study of ecosystems and has thus helped to understand their productivity.

From the first standpoint, there is no doubt that knowledge is now more comprehensive and systematic; it has covered subjects insufficiently studied before the IBP (e.g. nitrogen fixation by blue-green algae and symbiotic fixation through leaves). Moreover, methods for an easier and more accurate evaluation of the efficiency of fixation have been introduced to general use (e.g. the acetylene assay technique). Of IBP publications, J. M. Vincent's Handbook no. 5 has been widely appreciated as a manual for practical study of root-nodule bacteria and the IBP Catalogue of *Rhizobium* strains by Allen, Hamatová & Skinner has established conditions for much wider use of *Rhizobium* for seed inoculation. This is most needed in the developing countries in connection with the protein-deficiency problem. Indeed, the use of symbiotic nitrogen fixation for the improvement of legume yields is so promising that it was hoped FAO would use its authority in its promotion. Until now, however, little has been done in that direction, and the developing countries have hardly profited directly from the new techniques. This is due mainly to the lack of well equipped laboratories and well trained investigators.

From the second standpoint, thanks to the more accessible and accurate techniques, many important quantitative results on the contribution of microbial nitrogen fixation to individual ecosystems have been presented; those relating to extreme conditions are of particular interest. None the less, IBP has taken only the first steps in the integration of nitrogen fixation with ecosystem studies. Therefore it is much to be hoped that this will be included in the MAB programme, building on the basis provided by IBP. So far, it has been difficult to raise sufficient interest in such practical exploitation of the results of PP-N research, a reason for this being, perhaps, that the methodology of microbial ecology is different from that of most other ecological studies.

In spite of these critical remarks, however, achievements have been substantial. For example, the probability has been revealed that the root zone of many plants which have not hitherto been thought of as nitrogen fixers, may contribute significantly, especially in tropical and sub-tropical areas. In aquatic and wetland systems, the contribution to the nitrogen economy of both natural and agricultural habitats made by blue-green algae also appears to be greater than expected. Studies in India on the production and application of inoculants of these algae are of particular interest.

The genetics of nitrogen fixation in plants and bacteria are now attracting much greater interest than some years ago, and this area of research is developing rapidly in conjunction with studies involving biochemical and tissue-culture techniques. Another factor which has been underlined in these researches is the significance of light (through photosynthesis) and moisture relationships on symbiotic nitrogen fixation by legumes. The impact of this work is likely to be seen in the modification of existing agronomic practices. Much the same can be said of non-legumes, although their contribution to the nitrogen economy of the world is mainly in non-agricultural habitats. However, we still need to know far more about how and to what extent the biological process of nitrogen fixation contributes to the balance and sustained efficiency of ecosystems.

For practical agriculture, the proper balance between the contribution of microbial nitrogen fixation and the direct application of mineral nitrogen is of importance. This problem has been studied under various conditions of agriculture in different climates, and clearly ought to be continued. A new area for the introduction of microbial nitrogen fixation, especially of the symbiotic type, is in the reclamation of soils devastated, for example, by open-cast mining, where legumes inoculated with rhizobia can be extremely useful.

Looking to the future, the co-ordinated study of biological nitrogen fixation ought to follow two lines:

(a) Deepening of the knowledge on the efficiency of the fixation process itself, and on its integration in the nitrogen cycle. The nitrogen transformations need study in the fixing organism as well as in the "consumer" of the nitrogen fixed.

(b) Integration of this research with the studies of ecosystems, which has well defined perspectives, ready to be pursued.

In conclusion, numerous scientists from all parts of the world took part in the activities of the PP section, its sub-committees and intersectional working groups. Many scientists devoted much of their time and effort to co-ordinating the programme or synthesising its results at both national and international levels. They deserve to be thanked for this work and for the spirit of friendship and mutual understanding which has been prominent. It is urged that the specialists who have been grouped around the PP section be enabled to continue their collaboration through some continuing agency.'

Conservation

Section CT was at the beginning of IBP in a situation different from other parts of the programme in that there was an international organisation already established for the purpose of worldwide conservation of nature and natural resources, namely the International Union for the Conservation of Nature and Natural Resources, which had already gained wide support and prestige. IUCN, although it had an Ecological Commission, was active more in the administrative, management and political fields than in the scientific, so CT set out to provide a better scientific basis for conservation of the world's ecosystems, recognising that at the end of IBP, IUCN would inherit its work. The principal tool devised for this purpose was the Check Sheet Survey of Areas, the synthesis of which forms a volume in this series. Meanwhile, some achievements and disappointments of IBP's conservation efforts are recorded below in a personal account by the International Convener, E. M. Nicholson.

'Only six weeks after the planning committee meeting at Morges in May 1962, CT got off the mark with an international working group at the First World Parks Conference in Seattle, USA, which defined its main task as: "The establishment of the necessary scientific basis for a comprehensive world programme of preservation and safeguarding of areas of biological or physiographical importance for future scientists", and went on to list types of both natural and semi-natural environments to be included, in the interests of research as well as of conservation. For instance, it was stressed that: "Full account should be taken of the areas of land demanded by increasing modern field experimental research, and of the probability that the nature of many experiments may modify the sites used for them, and thus render necessary the provision of additional areas for future studies." It was added that "If the Programme includes a number of sites at which repeated measurements of biological productivity are made, such sites would require preservation for future comparison." It is regrettable to have to record that despite repeated attempts to secure notification to the data bank of basic information concerning important sites used for research actually within IBP, the programme ended without this having been accomplished in too many cases.

During the ensuing year, still before ICSU had formally agreed to set up SCIBP, preparatory steps were taken, including a full-scale field trial during the Jordan Expedition of April 1963 which led to the publication a year later of *An approach to the rapid description and mapping of biological habitats* by Poore & Robertson (1964). This was followed in July 1964 at the First General Assembly in Paris by a meeting to review the most promising systems available for use as a basis for the classification, and the systematic inventory, of the earth's ecological types. The long competitive struggle to establish a generally recognised "best" classification of ecological types had unhappily led to a partitioning of much of the world among schismatic schools each devoted to a particular model, often championed by one leading personality in an emotional atmosphere. As one very distinguished authority put it, to withdraw support from a system and its protagonist after being closely involved in it was no less painful and traumatic than parting from a mistress after an intense love affair.

It was felt in CT that there must be some more objective and scientific

81

method of handling the practical problem of marshalling and retrieving data. In principle there might well be several "best" classifications to serve different functions. For example, in mapping vegetation there could be as many as four best systems according to the scale adopted. Within a particular country or region a system tailored to its peculiar range of climates, geomorphology and plants and animals could show advantages over a worldwide system, provided the sacrifice in international comparability was acceptable. Again, some systems were so demanding that they could only be used accurately in the field by a handful of professional botanists, while even understanding their results required specialist knowledge.

Such thoughts led eventually to the idea of a "classification of classifications" according to their practical applicability and to the usability of their results. Such an approach promised both to assist the choice of a most appropriate system for a particular purpose and to stimulate more constructive evolution of the various systems through comparative rating of their merits and demerits, in an atmosphere more conducive to dispassionate appraisal. The system developed by Ellenberg in response to CT's original request became linked with UNESCO and its quite distinct requirements for vegetation mapping, and it became clear that to await its finalisation would be irreconcilable with the timetable to which IBP was committed.

Fortunately for CT, F. R. Fosberg had long been working on a simple and purely physiognomic classification of plant formations which had three great merits. It avoided all mixed categories blending climate, vegetation and other elements. It was based at least as much on tropical as on temperate experience, and had been widely tested in many different regions and situations. Finally, it did not essentially require capacity for exact identifications. At a technical meeting in 1966 it was accordingly decided to adopt the Fosberg system not as a 'classification' but rather as an inventory of plant formations. The decision to resort to computer data processing methods for storage, retrieval and analysis of results made such a system indispensable, and no other which could meet the worldwide requirement was available for immediate use. In some quarters the decision was misunderstood as prejudging the adoption of a worldwide vegetation classification – an entirely different matter with which CT was not concerned. On the contrary, it was stressed throughout that such classifications are in some ways analogous to languages, where global uniformity is impossible as well as undesirable.

The real problem is to secure the maximum translatability between one and another, backed by a strong group of, as it were, linguists in classification who know how to handle the problems and nuances of comparative analysis of data prepared according to different systems favoured by different schools. Valuable discussions were held, for example, with exponents of the Montpellier, Scandinavian, Holdridge and Kuchler methods, and it was made clear that in any territory where a particular system is well established and understood, survey material would be acceptable on that basis, provided that help was given in working out means for translating it and reconciling it with the international CT standard inventory.

Throughout this period intensive work had been in progress on the development of the basic check sheet for survey of areas. Mark I had been devised as early as August 1964, and by 1965 Mark III was ready for field trials in various parts of the world, including North and South America, Europe, Asia and Antarctica. In the light of these trials, and of a fuller assessment of total requirements, the check sheet was remodelled in 1966, leading through successive revisions to the final Mark VII version used in the main operation from 1967 onwards. There were inherent conflicts in devising a form sufficiently clear and simple to be used by many untrained surveyors and to be translated, as it was, into French, Spanish, German, Polish, Czech, Serbo-Croat and Russian.

The resulting data had to be adequate for a range of scientific analyses, although originally it had been planned to supplement them by large samples dealing in much greater depth with vegetation, geomorphology, land use and animal ecology. Unhappily the procedures and delays in two-way communication almost nullified this important second stage, which one can only hope will be resumed in the future by specialist teams from the appropriate fields commanding the necessary expertise.

In considering the scope and usefulness of the existing data bank it must be emphasised that its content has at this stage been restricted to a relatively small and elementary part of the data which it was conceived to store and make available for analysis and worldwide use. What should by now have been a major operational tool for sound decisions on worldwide land use has, owing to lack of financial and input support, emerged as so far no more than an interesting and sophisticated demonstration that the task which science and physical planning needs to be done can be done as soon as there are adequate resources. If and when that stage is reached the development of the computer data bank by CT will save some years in obtaining usable results.

Having commented adversely on the constraints it must be emphasised that in the comparatively few countries which were quick to set up strong and well supported CT sub-committees some very substantial and highly valuable results were obtained, not only for international but for effective national use. Canada, with the aid of an enviable financial allocation and vigorous leadership, both at federal and provincial level, demonstrated not only that good organisation could produce good returns in quantity but the surveyed areas could be demarcated on official maps on sale to the public, and could even be carried forward to statutory protected status under new legislation. This has already occurred, for example, in British Columbia. Impressive results were also obtained in countries as widely spread about the world as Australia, Japan, South Africa, Tunisia and the Netherlands. In some cases special attention was given to the extent of coverage of plant communities and species afforded by existing protected areas; in others to the better understanding of the nature and variants of such communities in an international context; and in others to the completion of a country check sheet summing up a series of returns for particular areas.

In the planning of the check sheet survey it had been hoped to secure valid returns to the data bank for something of the order of 10 000 areas. The acceptable minimum was set at 3000, which in the end proved to be all that could be secured, although it is known that a substantial additional number got trapped somewhere in transit from the surveyors and never reached the data bank. On the whole the global distribution of returns was rather more satisfactory than the grand total, although, naturally, returns from the tropics and from the Third World in general were much less than would have been obtained had the original expectations materialised for international funding and dispatch of binational or multinational survey expeditions to critical areas.

In addition to participation in the check sheet survey some 43 countries initiated projects of their own within CT, to a total of 276 reported, of which some 185 in 36 countries had either been successfully completed or were still on the active list at the conclusion of the IBP, most of these having been assessed as ongoing long-term studies. The diversity of subjects and even of languages involved, and the scale and quality of the projects, defeat any attempt to generalise about them at this stage; but in view of the peripheral role of CT in terms of its research as distinct from its survey component, it must be considered satisfactory that so many countries found it worthwhile to conduct studies within this section. That is perhaps a good augury for the gradual build-up of professional research in ecology applied to conservation, and consequently for bridging the gulf between these subjects.

Among other important concepts of CT was the build-up and evolution of international and national field stations to form eventually a worldwide network of biome research stations, for which such initiatives as the Serengeti Research Station in Tanzania, the Charles Darwin Station in Galapagos and the Azraq Field Station in Jordan were envisaged as prototypes. Unfortunately, the Azraq Station was early engulfed by hostilities in the Middle East, and although held intact could not be restored to a functioning role before the end of IBP. Enquiries produced a wealth of material confirming the high potential of such stations but they are not yet adequately appreciated or supported. In view of the interest of IUCN in the growth of such a series of strong-points for world ecology, proposals were worked out and agreed with a number of leading directors for a measure of liaison and communication to be established under IUCN auspices, but full implementation of this idea has not yet proved possible. The best that can be said is that most of these scattered and remote stations are barely surviving, with setbacks from which even the strongest are not yet immune, but that a growing number of young biologists are proving keen enough to accept the hardships and sacrifices of serving in them for a spell in order to acquire a unique type of ecological experience while it can still be had. The organisation of much of the US IBP on a biome basis, and the series of biome volumes resulting, will no doubt help to spread the concept and to build up further support.

The CT section also took a number of other initiatives. One of the most important, in response to a request from the 11th Pacific Science Congress, was the publication of a List of Pacific Islands with special reference to their conservation opportunities and problems, which led to the drafting by IUCN of an international convention on the conservation of islands for science. Another was the convening, in conjunction with the Rome Assembly of IBP in September 1970, of a working group on environmental considerations in development projects, with the participation of IUCN, the Conservation Foundation, the World Bank, the Ford Foundation and other interested parties, including FAO who were the hosts. This led later to the publication of the book by Dasmann, Milton & Freeman (1973), which digests worldwide experience concerning the ecological problems arising in the course of development projects up to 1971.

During its brief existence CT found itself cast as a kind of Janus, looking hopefully towards the serried ranks of biologists and hoping to persuade them of the significance and future promise of ecology as applied in conservation, while at the same time looking appealingly towards conservationists, including such professions as foresters, to urge that their daily tasks would gain in status and effectiveness if they could make more imaginative use of the background and foundation offered to them by the rapidly advancing life and earth sciences within their evolving environmental framework. The role was a difficult one, and it often seemed that intense effort yielded no more than scanty results. But the task is essentially long-dated, and perhaps in retrospect it may prove that substantial movement has occurred.

By arrangement with IUCN it falls to the International Union to develop and use the world data bank and its methodologies so far as they justify themselves in the future circumstances of environmental conservation. In a sense CT represented a temporary and carefully controlled breakaway from IUCN, to provide it with a comprehensive scientific foundation and equipment which had been implicit in IUCN's objectives, but which, owing to its exacting demands and formidable scale, had proved impossible for IUCN to tackle in the course of its everyday duties. Most of those who guided CT were loyal and well tried participants in IUCN's successful struggles, and were ever conscious of the difficulty of underpinning the opportunist practical defence of ecosystems and threatened species with the fruits of fundamental research for which it could not wait. If world conservation grows steadily more scientific, and if the biological and earth sciences grow steadily more conservation-minded, the work of CT will not have been in vain.'

Plant genetic resources

This theme of IBP was part of the Section UM, other major parts of which were concerned with biological control of selected pests, and with food protein resources. Synthesis volumes on each of these subjects are due for publication soon. UM is usually listed as the last of the seven sections, but, as an example of the work, plant genetic resources fit conveniently here, with other branches of terrestrial biology. The following summary is by the theme's co-ordinator, who was also a Vice-President of SCIBP, Sir Otto Frankel.

'There are many references in this book to co-operation between the

IBP and the specialist agencies of the United Nations, particularly UNESCO and its MAB Programme. IBP's work on "plant genetic resources" – a phrase which was coined during the programme – is an example of close co-operation with another of the agencies, FAO, and led, within the decade, to a wide recognition of the importance of this subject. The central contribution of the IBP in this case was one of ideas and initiative, ending in significant contributions to knowledge and resulting, through chain reactions, in international action.

In the original programme (see page 45) a survey of available knowledge and genetic variation of major crop plants and their wild relatives was included, with emphasis placed on information and availability, rather than on preservation in the face of impending loss. It was originally envisaged that as in most of the IBP, the work would be organised by committees and groups in participating countries, and indeed there were a number of worthwhile national contributions. However, outside the few institutions, mostly in developed countries involved in exploration and conservation, there was, to begin with, little appreciation and almost complete lack of financial and logistic support. Moreover, it became evident that there was an urgent need for the clarification and development of the scientific methodological infrastructure, and in these areas there was an obvious role for IBP.

A transformation of ideas came however in the mid-1960s from a variety of sources. Warnings of a greatly accelerated rate of displacement of primitive crop varieties by locally selected or introduced cultivars came from scientists with experience in important gene centres, notably from Iran, Turkey, Ethiopia, the Near East and Latin America. At the same time there was a rapidly growing interest among plant breeders in the primitive cultivars and the immense range of variation they contain, and in the wild relatives of economic species, many as yet scarcely explored or exploited. Stimulated by developments of this kind, FAO initiated a review of its activities and responsibilities in plant exploration and introduction, and Sir Otto Frankel, leader of IBP's group on plant gene pools, was invited to carry out this review. As a result, a working partnership, which had been inaugurated at an IBP meeting at FAO headquarters in 1965, led to a virtual integration of effort which continued until the end of IBP. There was a technical conference prepared jointly by FAO and IBP in 1967; IBP handbook no. 11 was largely the outcome of this conference. Nearly every year there were meetings of FAO's Panel on Plant Exploration and Introduction, which included members of IBP's working group, and these meetings, together with other frequent contacts assisted FAO in planning its programme on genetic resources.

Thus the scientific background of the subject was laid down, the methodologies of plant exploration and genetic conservation were examined and clarified, goals and strategies were defined. The multi-disciplinary nature of the problems was recognised by specialists in many of the relevant fields, from ecology, genetics and systematics, to plant breeding, agronomy and computer science. The primitive varieties of traditional agriculture were designated as primary targets for urgent exploration, with wild relatives, especially where threatened, as secondary targets. A survey of genetic resources in the field was to supply required information on priorities for exploration. Long-term conservation of seeds, where applicable, was recognised as the safest and most economical method of conservation, and a co-operative network of seed storage laboratories was seen as the most efficient and economical form of organisation. The documentation of genetic resources was clearly essential if they were to be put to practical use in research and in breeding practice. Last but not least, co-ordination by FAO was regarded as essential for planning and executing an action programme with broad international participation.

During the most recent years a good start was made in carrying out a programme on these general lines, particularly in the survey of genetic resources in the field which was carried out by specialists in many countries in 1972, jointly financed, and published in 1973. There was an obvious and close relationship between the conservation of genetic resources on the one hand through seed banks and other forms of living collections, and on the other hand their survival in natural habitats such as nature reserves and national parks, and this brought the Plant Gene Pools Theme of IBP in close touch with its Section CT. It is satisfactory to see that the MAB project "Conservation of natural areas and of the genetic material they contain" includes among its objectives the long-term preservation of wild relatives of domesticated plants and of wild plants directly useful to man, thus continuing and expanding the concern for the preservation of the broad range of genetic resources which IBP had stimulated.

The urgency of the situation has been brought home to scientists, administrators and politicians, and in this also IBP played a part and became as it were a catalyst for action. Indeed, the prominent place which the conservation and development of genetic resources took at the UN Conference on the Environment at Stockholm in 1972, and the subsequent elaboration of UNEP's plans are due in large part to IBP initiative. The synthesis volume entitled *Crop Genetic Resources for Today and Tomorrow*, no. 2 in this series, marks the end of IBP's contribution to this subject, but also, it is hoped, the opening of a new era of active work.'

Wetlands

The term 'Wetlands' has been used for purposes of IBP to cover shallow-water areas which may be fresh, brackish or salt, but which are usually characterised by having emergent vegetation. As such, the wetlands provide a link between terrestrial ecosystems on the one hand and aquatic ecosystems on the other.

Certain of the wetlands which were subject to productivity studies under IBP, for instance the delta area of the Danube with its massive reed beds, are among the most highly productive ecosystems known anywhere. One reason for this is, of course, that water is not a limiting factor and that nutrients in some quantity are brought down in silt. Such areas, and even more so the larger areas of papyrus and reed swamp in the tropics, such as the Sudd region of the Upper Nile, are the cause of very large water loss through the transpiration stream, but at the same time retard the flow and influence the quality of the water which passes through them, so they are of great economic importance in relation to water management.

In the IBP, the main interest in wetlands has been in their primary production, but also to some extent in other constituents of energy flow, in their capacity for storing certain amounts of mineral nutrients, in their structure in relation to natural environment, as well as their management. As in other parts of the IBP, there was a bias towards the northern temperate zone, due to the relatively strong participation of its countries. But the scientists involved are willing to extend the scope of their activity in the research, conservation and management aspects of wetland ecology, especially to the tropics and sub-tropics.

Wetlands provide a good example of intersectional activity combining the interests of all sections of IBP, but the need for creating a special working group was felt rather late in the programme, at the IHD/IBP meeting on the Productivity and Ecology of Aquatic Macrophytes in Bucharest and Tulcea in 1970. This meeting had been organised by Romanian and Czechoslovak scientists with the assistance of the PF and PP sections, and later with the assistance of Polish specialists a working group was formally established at a meeting at Mikołajki in 1972. The group's activities have been serviced by the PP secretariat in Prague, which is supervising the preparation of a wetlands volume of synthesis.

Floating water-weeds are characteristic of some wetlands and, like rooted vegetation, are of great importance in the hydrological regime. Great trouble has resulted from population explosions when water plants are introduced from one region of the world to another, as in the case of the water hyacinth (*Eichhornia crassipes*), in many parts of tropical Africa and Asia, and the water fern of Lake Kariba (*Salvinia molesta*), both of which are indigenous to the New World but not the old. These and other water-weed problems have been studied by IBP in co-operation with IHD.

The conservation of wetlands is being greatly stimulated by the International Convention on Wetlands of International Importance which was signed at Ramsar in Iran in 1971, but the definition of wetlands is different from the IBP usage: 'For the purpose of this Convention wetlands are areas of marsh, fen, peatland or water, whether natural or artificial, permanent or temporary, with water that is static or flowing, fresh, brackish or salt, including areas of marine water the depth of which at low tide does not exceed six metres.' This convention arose from the consideration especially of waterfowl habitat, concerning which the International Waterfowl Research Bureau had been particularly active over a number of years. The definition of wetlands was drawn extremely widely, and from the viewpoint of conservation of areas this is convenient, since lakes, rivers and areas of shallow sea are covered without the need for more conventions. However, when describing biomes or ecosystems, and particularly when considering their biological productivity, there are fundamental differences between deep lakes, shallow lakes, rivers, streams, shallow seas, salt marshes, etc.

Inland waters

The following account of some of the highlights in IBP's work is based on notes prepared by the section PF scientific co-ordinator, J. Rzóska, following a final meeting of the section's leaders with its Convener, Livia Tonolli, held at her institute at Pallanza on Lake Maggiore. In its research work the PF section moved from general 'concepts' to an increasingly realistic approach to the factors controlling biological production in fresh waters. From the beginning the necessity of collating methods was recognised and these were worked out in a series of specialist meetings, starting in 1964, and resulting in publication of the series of five handbooks which cover all levels of biological productivity in fresh waters and the main factors which control it. Several of these books were published too late to influence the choice of methods in IBP itself, but they are now in use by many limnologists.

PF's work actually started before the formal opening of the programme by an appeal for co-operation to limnologists in 1963, which on the whole met with favourable response. The focus on production problems was attractive, and in the years following, a network of research sites emerged, with research financed through national committees. These sites represented, though unevenly, the major geographical and climatic regions of the continents.

The organisation and execution of research at the many sites proceeded unequally, for in some countries teams had to be created from scratch for investigating the many factors composing an aquatic body. A number of younger scientists were offered opportunities which did not exist before. In other countries, with established traditions of production studies, a considerable stimulus was created for expansion of research at pre-existing sites. This has been particularly evident in the USSR.

Among major sites, climatically important, which would not have been investigated intensively without the stimulus of IBP/PF are Lake Char, far north in Canada, Loch Leven in Scotland, the tropical Lake George in Uganda, and Tasek Bera in Malaysia. It is also significant that a number of countries which showed little or no response to the first appeals for co-operation later joined and produced significant research results.

The PF office in London, acting as information centre for enquiries on research, scientists involved, and bibliographic references, served as liaison centre. It also provided fruitful help for the work of the International Hydrological Decade of UNESCO. This was especially so in the field of excessive water vegetation, a major economic problem especially in tropical and sub-tropical countries.

The main research results which provide the material for synthesis consist of some ninety unpublished 'Data Reports' and hundreds of published papers. These were subjected to statistical analysis, modelling, and group discussion of biological components. Some general points which contribute to a better understanding of the working of aquatic ecosystems and which have emerged from the assessment of PF results are as follows.

The statistical analysis of ninety lakes is 'new' because the lakes are from different climatic regions. Correlations found by analysis are therefore of wider significance than previous attempts which have dealt mainly with lakes of one region. Modelling the total production process, which is based on a less extensive range of lakes and parameters, may enable some predictive conclusions for management, though new parameter values are needed to make prediction sounder. Dealing with the three big components of biological production – (i) primary, (ii) secondary including fish, and (iii) bacterial production and decomposition – the IBP work demonstrated that there are still considerable deficiencies in our understanding of the functioning of the aquatic ecosystem. Some of the outstanding problems are: difficulties in assessing primary production exactly, largely because of respiration assessment; in secondary production, changes in trophic habits by many animal components; the role of bacteria as food for some organisms, and that of allochthonous matter, require much further attention; the values of particular nutrients is still controversial and so also is their recirculation. Being well aware of these and other gaps in knowledge, the value of PF results will also provide guidelines for future research.

In the conservation of scientifically important areas and sites around the world, the PF section performed a major task in Project Aqua. This consisted of listing, after extensive consultation which included CT and other sections of IBP, some hundreds of sites of a kind which merit some measure of protection or conservation. They range from large lakes and stretches of river to quite small sites which in some way are unique or which have received an important investment of scientific research and have therefore become reference points in limnology. For each site basic data are presented in a uniform way, with the reasons for its protection, the risks to which it may be subject, and appropriate references to literature. This project was in fact started by SIL before the existence of IBP, but the IBP world network of hydrobiologists coupled with efficient organisation of the PF section was the major influence in driving it forward, through a provisional draft list issued in 1969, to the more comprehensive list published in 1971 as IBP handbook no. 21. At the end of IBP this was revised with many important additions, and the project, like the terrestrial conservation work of Section CT, has been handed over to the IUCN. It is intended in due course, to bring Project Aqua, together with the Project Mar which is a comparable list of shallow waters and swamps of importance to aquatic birds, and lists of areas of other wet ecosystems such as bogs, salt marshes and

Plate 1. Ngorongoro Crater in Tanzania, 250 km² of highly productive grassland, grazed by up to 24 000 large wild herbivores of a dozen species. This area and its fauna were studied by the Serengeti Research Unit.

Plate 2. A Tundra biome study site: a view of Pronchitscheva Bay, USSR.

Plate 3. A corner of Pasoh Forest in Malaya. This is the site of a major research project which was arranged jointly by the IBP committees of Malaya, Japan and the UK.

Plate 4. Snares Island, south of New Zealand. This semi-arctic island at latitude 48°S. was a major IBP site studied intensively by New Zealand biologists. Like many islands it has a high proportion of unique flora and fauna. Photograph: J. Warham, Christchurch, N.Z.

Plate 5. Inechua men of Nuñoa, Peru (altitude 4300 m) pack their llamas to go to market. The Inechua were studied as part of the HA work on high-altitude peoples.

Plate 6. Lake George in Uganda. Fishermen with equipment and catch in 1930. Subsequently this shallow lake of about 250 km² was developed as one of the world's most productive natural fisheries (per unit area).

Plate 7. The IBP research centre on Lake George; members of the research team taking observations on the lake. The measurement and analysis of production factors in this lake was conducted jointly by the United Kingdom and Uganda over the five years of IBP Phase II.

shallow seas, into a comprehensive directory which will conform to the definition of wetlands quoted on page 90 from the International Convention. This comprehensive list will then become a kind of blueprint for the creation of nature reserves, national parks and other protected areas.

With hindsight it can now be suggested how the results of PF could have been improved. Firstly, recommended methods of research should have been clearly defined at the beginning so as to make results more comparable; this applies also to units of quantitative assessments and chosen parameters. Secondly, a better selection of research sites should have been adopted to reflect the variety of conditions existing in inland waters, which have been described aptly as 'islands of water in a sea of land'. Thirdly, these 'islands' of water are greatly influenced by their surrounding drainage basins, which indicates that the work of PF could with advantage have been integrated more closely with that of PT.

The seas

Max Dunbar, the International Convener of Section PM has contributed the following summary.

'Before the IBP started, international co-operation in marine research, at least in the North Atlantic region, was already well established since 1902 with the formation of the International Council for the Exploration of the Sea (ICES) having headquarters in Copenhagen. The international interests in the Antarctic whaling industry had also established certain co-operation in research and exploitation, and more recently, fishery research and management in the North Atlantic came under international discussion and supervision with the formation of the International Commission for the North Atlantic Fisheries (ICNAF). Co-operation had also been established in other regional areas, such as the Caribbean and the Indian Ocean. Moreover, within the ICSU family, SCOR had been active for a number of years and SCAR was concerned with several aspects of marine biology. There was also the IOC responsible to UNESCO and a measure of international co-ordination of fisheries research organised by FAO.

Nevertheless, there was clearly valuable work to be done in the sea through the means developed by IBP, especially in coastal waters and with reference to fields of activity, and to specific needs, which had not been attacked internationally, notably in aquaculture, marine pollution, and the study of the factors determining marine productivity. These three formed the subject matter of the first three themes developed by

PM. A later addition was the study of the marine mammals, especially the seals; but the sea mammal sub-section did not develop a substantial programme of its own partly because other organisations were involved in this subject. A theme on coral reefs was also started under the stimulus of Sir Maurice Yonge, who was scientific co-ordinator of the section for several years, and held several meetings including participation with CT in the important meeting at Koror in 1968, and joint sponsorship of the international symposium on coral reefs held in Australia in 1973. A plankton statistical project, launched early in the programme and continued until 1971, was based in New Zealand and had the objective of preparing research and review papers which would ultimately form the nucleus of statistical methods as applied to plankton research, both phytoplankton and zooplankton.

Working groups were established early on benthic and littoral communities, brackish and turbid waters, and intercomparison of methods. The section also participated jointly with SCOR in working groups on the estimation of primary production in the sea; on continuous monitoring in biological oceanography; and on phytoplankton methods. Methods for the study of benthos particularly needed examination and became the subject of IBP handbook no. 16.

The activities of the section as a whole were brought into closer focus and given new impetus at a working conference in Rome, September –October 1971. At that time 300 separate projects carried out by thirty countries figured on the PM list, but many of these had already been completed. Of the rest, some ninety projects were found to be truly viable at that time, and these provided the basis for a revised plan of action which included primary production, secondary production, modelling, marine environmental damage, aquaculture and higher trophic levels.

Important discussions were held also at the conference on whales held in Virginia in June 1971 on the initiative of the US Marine Mammal Council, at the SCOR Sea-Ice Conference in Iceland in 1971, the Indian Ocean Expedition meeting at Kiel in March 1971, and the Conference on Seals held at Guelph, Ontario, in August 1972. A specifically IBP symposium was organised at the final International Congress of Zoology at Monaco in September 1972, on the biological effects of inter-ocean canals; and another in June 1973 at Malta on problems of the eastern Mediterranean. This last had been planned in the early years of IBP but had been postponed repeatedly for various reasons, including the death of Gunnar Thorson, one of the prime movers in the project. The Malta symposium under Neil Hulings resulted in a plan of action for the eastern

Mediterranean, which attracted the attention of several international bodies and which, it is hoped, will form the basis for ongoing attention and action on the serious conditions now existing biologically in that area. Finally, a conference on the grey mullet was organised and realised by O. H. Oren in Haifa in June 1974, the culmination of a great deal of planning.

Many hundreds of papers in scientific journals have resulted from PM projects, and significant advances are to be credited. Some indication about them emerges from a brief review of national participation. Soviet contributions include important work on trophic relationships and the functioning of marine ecosystems, particularly in pelagic systems in tropical seas, but also in benthic systems in the coastal regions of the USSR, and on the vertical distribution of phytoplankton in the open ocean. Canadian contributions include: (*a*) the interesting discovery that the primary production by attached algae in inland seas may be three times as high as the production of phytoplankton; (*b*) a year-round study of ice biota and nutrient levels in ice in Frobisher Bay, Baffin Bay; (*c*) the measurement of primary and secondary production in the Gulf of St Lawrence over a period of four years; and (*d*) the factors governing production in the Strait of Georgia, British Columbia. In Japan the work included most useful studies of the productivity cycles in the Seto Inland Sea, in Sendai Bay and in the waters of northern Japan; also in the littoral regions of the Kuroshio Current area. France undertook an elaborate and highly successful study of the ecology of coral formations in the south-west Indian Ocean, and on the estuarine effects on productivity of the region off the mouth of the Rhone. The Argentine contributed work on the marine algae of the Argentine coast, a study of primary productivity in the South Atlantic and Antarctic, and of the influence of bacteria on the estimation of primary production. The United Kingdom programme was elaborate and varied, including work on marine food chains, on the pollution of marine ecosystems in Britain, on marine fish culture, plankton distribution in the North Atlantic, oceanic ecosystems, benthic communities, and invertebrate larvae. The *Marine Pollution Bulletin,* begun as an IBP project under R. B. Clark, was taken over by the publishers Macmillan and is now a regular monthly journal. The United States carried out a study of upwelling systems and the mechanism of upwelling in general, on marine mammals in the Bering Sea, and on marine pollution. The work done in Romania on the production of microphytobenthos emphasised the unrecognised importance of this element in the system, and so did the work in the

Wadden Sea carried out by the Netherlands. Norway contributed work on phytoplankton in sheltered fjords and on pollution in the Oslofjord. South Africa put a lot of its marine research under the IBP label, so that the results are varied and abundant, mainly on coastal productivity. India did excellent work on inshore production, littoral ecology, and seaweed culture. The Philippines specialised in work on seaweeds and seaweed production. Italy concentrated on the northern Adriatic, general and biological oceanography, and on certain aspects of aquaculture. The New Zealand programme produced results in benthic communities, including algal populations and *Spartina* beds, on phytoplankton production and on statistical plankton study.'

The following quotation from J. H. Steele at the Rome 1971 conference serves well to conclude this brief description of the PM effort:

'I cannot summarise discussions whose complexity has matched the subject, but there are three points where we appear to have common ground. Firstly, there is a need for much greater interaction with studies of terrestrial ecology, not merely by comparison of biomass or energy flow but through evaluation of possible similarities in the factors that control such systems. Secondly, we have accepted the idea that simulation modelling can be useful not only as a general approach but especially as a process deeply involved and integrated with the research processes of observation and experimentation. We might even be too enthusiastic at the moment but such detailed involvement is likely to lead to a more balanced appreciation of the value of such theoretical models and, in turn, to the theoreticians acquiring a greater insight into the kind of models that are needed. Lastly, we have begun to see how this combined approach involving observation, experimentation and theory can be applied to practical problems. This is particularly true for pollution where the intensive study of inshore marine ecosystems leads naturally to the application of the results to deterioration in such regions.'

Mankind

The HA section got off to an early start, even before the IBP had officially begun as described by its International Convener, J. S. Weiner, in Chapter 1. By the end of Phase I its team of theme leaders was well established, work was proceeding on agreed methods of research, and many participating countries had either already commenced research projects or were about to do so. Dr Weiner has contributed an account of its subsequent activities, which follows.

'At the inception of the operational phase the HA section was assured of strong international support. It was already clear that major themes were emerging such as the human adaptability of circumpolar and high-altitude regions which it must not be forgotten represent potential areas for the future settlement of expanding human populations; their ecological peculiarities deserve the closest attention both on scientific and practical grounds. Also apparent was the considerable interest in using heights of children and the level of work capacity as indices of response of human populations to a large variety of ecological circumstances.

In gaining acceptance of the HA "idea", a series of regional meetings were organised, many of these also serving as forums for the discussion of methodological questions. As at earlier meetings – the CIBA Foundation in 1962, Burg Wartenstein and Paris 1964 – these regional meetings brought together, often for the first time in a working relationship, demographers, physiologists, geneticists, nutritionists and other medical scientists. They took place in Japan, Poland, India, Norway, Canada, and the USA. The Convener was also able to meet groups in Budapest, Geneva, Prague, South Africa, Holland and the USSR, as the operational phase got under way.

Parallel with the building up of the programme through national committees and correspondents in some fifty countries, the major task of compiling a handbook of agreed procedures for field work went on. This was no simple matter of combining the literature and correspondence with recognised experts. To get full agreement on some of the more important and to some extent controversial techniques it was necessary to hold many discussion groups of experts. Such meetings were held in Norway, the UK, Japan, the USA and Canada, and Geneva. It became inescapably necessary to arrange for validation trials and the comparison of techniques. Co-operative methodology exercises of this kind were carried out on a multinational basis in Toronto to decide on methods of assessing work capacity. In Cincinatti another large team examined different procedures for assessing heat tolerance. Anthropometry was the subject of close consultation between HA and WHO experts. Individual laboratories undertook the evaluation of various procedures, for example of skin colour by reflectance spectrophotometry (Oxford), of demographic surveys (Paris), of blood sampling and collection in the field (UK), of cold tolerance (USA and Japan), of some aspects of ergometry (Holland and Norway), and many more.

It is true to say that every one of the fifty techniques described in the

IBP Handbook, no. 9 was the outcome of collaborative consideration and exchange of views between established authorities. IUPS, with the stimulus of its President Wallace Fenn, made a notable contribution by bringing into being and providing funds for a Commission of Physiological Anthropometry which clarified many of the field techniques. When teams moved into the field to investigate the biological conditions of communities it became apparent that certain features of a psychological nature could, in many cases also be fruitfully investigated. Accordingly, with the support of the International Union of Psychological Sciences, a study group of specialists met at the CIBA Foundation and produced IBP handbook no. 10. These two HA handbooks were not published until the operational phase was under way, but all the material of no. 9 was made widely available in duplicated form at the start of the operational phase.

Let us for a moment leave on one side the bald statistics of the project activities during the operational phase, and try to provide a picture of the kind of research that the human biologists were engaged on. For illustration we might look at the years 1969 and 1970 in the middle of the phase.

At this time HA teams were at work in the circumpolar regions of the three continents studying in detail the biology of environments of Eskimos, Aleutians, Lapps and Ainu. In ecological contrast other teams were active in the hot, dry centres of sub-Saharan Africa, in Jordan and Israel, the Karakoram Desert and Central Australia. Others again were in the hot humid regions of the Congo, SE Asia and the Amazon and along the East African coast. And there were teams in the highest reaches of the Pamirs, the Caucasus, the Simien Mountains of Ethiopia, the Himalayas and the Andes. Small communities living in a high degree of isolation on islands of Polynesia and Melanesia, of Tristan da Cunha and off Japan, in Yugoslavia, in remote lakes or marshes on the Danube, and in White Russia and Siberia, were being visited by IBP expeditions.

In all of these the health and fitness and nutritional condition of the communities were assessed. Particular attention was paid to the developmental status of the children to the physique of adults, to their working capacity and respiratory efficiency. Of special interest were the responses evolved by the peculiarities of the environment. Thus the cold tolerance of cold-climate peoples, the heat tolerance and water balance of hot-climate groups, the haematological and cardio-respiratory adjustments at high altitude, were all the subject of study. The demographic and genetic communities were not overlooked, especially among the many small and inbreeding groups living in conditions of isolation and difficult climate and terrain. There were also investigations under way into the biomedical effects of migration from rural to industrialised conditions, for example North African and Yemenite settlers in Israel, and the Polynesian islanders into New Zealand, and Bantu tribes moving into industrialised cities of South Africa. Comparisons between rural and urban communities were the subject of long-term study in several countries – in Czechoslovakia, Poland and the USA.

Throughout the operational phase, meetings of HA workers concerned with a common theme were arranged. Most highly organised was the series arranged by the circumpolar workers of the large Scandinavian contingent (from Sweden, Norway, Denmark and Finland) drawing in the Russians, who were also working on Lapplanders, the Japanese Ainu teams and the American, Canadian, French and Danish working on Eskimos of Greenland and Alaska and in the Aleutians. There were connections between these Arctic investigators and those in the South Polar Regions working through SCAR, itself a legacy of the IGY. This co-operation was a model of its kind, and in achieving it a leading role was taken by Sandy Hart of Canada, whose death in 1973 was felt keenly in the IBP as in many other circles.

Likewise the high-altitude groups achieved an effective measure of co-ordination through the work of the theme leader, Paul Baker, who organised an important meeting with WHO, and particularly through his personal visits to India and the Soviet Union, as well as unremitting work in the Andes. Some members of his Andes team worked also in other high-altitude regions.

The work of all sections of IBP through the whole decade, but especially during the five years of operations was hampered by two main sources of difficulty, namely shortage of money which was ever present, and political problems which were intermittent. Lest anyone should get the impression that IBP had an easy passage, it is fitting to end this chapter with the HA section's experience in these two respects.

First concerning finance: on the international plane and within the limits of their restricted budgets two organisations proved most helpful to the HA section, the Wenner-Gren Foundation and WHO. The former supported a number of conferences, notably the crucial 1964 meeting at Burg Wartenstein already mentioned, and it also gave grants to individual investigators. Within WHO the sections on children's development, genetics, cardiovascular, nutrition and demography were keenly interested in the IBP and in addition to the symposia arranged in collaboration with HA scientists, WHO gave research grants for HA work in India, West Africa, Congo, Israel, Polynesia. WHO was not, of course, the only specialist agency of UN whose interests impinged on the HA section. Another was FAO, and it came as a surprise that despite FAO's concern with the nutritional aspects of the programme no financial help was ever forthcoming. Later it appeared that the constitution of FAO precluded the kind of support that WHO was able to offer to HA and UNESCO to other branches of IBP. The HA section itself did not at any time approach any of the great international research foundations, but why these agencies should not have supported, even modestly, the work of SCIBP in any of its branches, working largely on a voluntary basis and to a considerable degree in the interests of the developing countries, remains something of a mystery.

On the scientific merits of the case, many national IBP committees were able to obtain funds from their national agencies for the work of the IBP over and above the ordinary allocations for scientific research. This falsified the early fears expressed in some quarters that IBP would cut into and compete with existing research commitments. As early as the formative meeting at Morges in 1962, Ledyard Stebbins predicted that the US Congress would in due course vote an annual grant of five million dollars for IBP work, and he was proved right. At the same time the USSR representative, Professor Steinberg (a man of great intellectual capacity and personal charm who died the following year) made plain that the USSR would back IBP strongly, and so it proved: during the operational phase the USSR HA committee was able to mount at least a hundred expeditions.

National support for IBP was in fact nearly worldwide; but in the developing sector lack of finance and of trained personnel inevitably reduced the scale of operations. HA projects prepared for India, SE Asia, Philippines, West, Central and East Africa, and several Latin American countries, could not be realised or only to a very limited degree. There was no lack of response on the part of the local physiologists, nutritionists and other medical scientists, but all too often not even a growth survey could be mounted for lack of resources. Nevertheless, a number of joint project partnerships between a scientifically developed and less developed country were effectively planned and carried out. Notable HA examples were projects between Brazil and the USA, Belgium and Zaire, Peru and the USA, Poland and Egypt, Tanzania and the UK, India and Belgium, France and Senegal, Malawi and the UK.

There were successful projects in countries which were unable to provide local resources; such were those in Chad (Belgium), Solomon Islands (USA), Tokelau Islands (New Zealand), Tristan da Cunha (UK), New Guinea (Australia), Bhutan (UK), and Zaire (Italy). Finally, there were many countries where it was possible to include ongoing research projects with the national IBP programme if they fulfilled the required criteria and methodology. This greatly strengthened the overall effort, and in the HA section it is probable that about 30 % of the projects came into this category.

Concerning politics, it is not surprising in our turbulent times that over the ten-year span of IBP troubles should have assailed and harassed scientists. There were ideological differences blocking movement into or out of certain countries – visa difficulties for East Germans to the West, for South Africans into the Eastern Bloc and to other parts of Africa, there was the impossibility of holding genuine multinational scientific meetings in South Africa, and the peculiar isolation of Cuba and mainland China. The Middle East conflict brought about the high-jacking of the Israeli HA Chairman and his detention for some months in Syria. The Biafran war in Nigeria undoubtedly restricted what would in peaceful times have been a substantial contribution. The outbreak of the first India–Pakistan war coincided with a major HA Conference in New Delhi. This went ahead but only one of the fourteen representatives from abroad was able to reach the conference, and this had a depressing effect on the Indian programme since bilateral arrangements for research were postponed and were not easy to achieve later on. In Greece the academic community experienced great political difficulties and the IBP contribution was on a minor scale. The saddest setback arose from the 1968 events in Czechoslovakia. The HA Chairman, the renowned physiologist Ottakar Poupa, who had effectively promoted the cause of IBP in the Eastern countries as well as in his own, was obliged to leave his native land. He was first and foremost a dedicated scientist, but also an outspoken defender of liberty.

Nevertheless, political conditions did not deflect scientists the world over from an endeavour which they regarded as beneficial for all mankind. In many of the countries experiencing disturbances field research was successfully carried out – in Nigeria, Zaire, Israel, Jordan, Egypt, Pakistan, and India. In the Soviet Union the fall of Lysenko made possible the incorporation of modern human genetics into many projects ranging from sero-genetic studies of population affinities to the consequences of inbreeding and consanguinity. It can be said without reservation that

101

within the sections of IBP, the scientists from the world over whenever they met, worked together and conferred together in the greatest amity, concentrating on their common scientific and humane goal without the intrusion of differences in political creed or ideology.'

5. Synthesis and transfer

A miscalculation was made at the outset of IBP. We assumed that the programme would wind itself up on its own, in a satisfactory way, at the end of the operational phase. This proved to be a fallacy, which was fortunately recognised in time to add a final two years in order to make adequate arrangements for synthesis of the results and transfer of on-going activities to permanent organisations. Among the many eminent biologists who have given much time and energy to the IBP, special tribute must be paid to the one who originally proposed that this should be arranged, namely Ivan Malek of Prague. Whereas in the first part of the programme his enthusiasm, drive and perspicacity were always with us, in the latter part his movements were so unfortunately curtailed that his contributions had to be mainly through deputies or letters. However, he continues as an important influence, especially in emphasising the all-pervading *processes* of productivity.

About half-way through the programme it became widely recognised, by national IBP committees as well as by SCIBP, that the results, whether reports on progress or final presentations, were becoming widely scattered, issued in many different forms, and there was considerable danger of some of them being lost sight of, or at least not being brought together in a co-ordinated form which would be of use to future generations of research workers and future programmes. Accordingly, discussions started on the problem of bringing results together into some form of synthesis, not with the object of having a memento of IBP, but as considered statements, containing guiding conclusions and drawing attention to the principle that the whole is greater than the sum of its parts.

It must be stated that the proposal from SCIBP to extend the programme from the eight years planned originally to a decade had its opponents among scientists in ICSU at the time. In the inner councils biologists were in the minority, and there was a serious danger that the extra two years would be disapproved. The scale was tilted, however, by the then President of IUBS, Don Farner.

Synthesis

Ideally an international synthesis of results should involve the data from all related projects being examined together, reworked and re-analysed

103

in appropriate cases, conclusions being drawn from the combined results. Furthermore, these combined results should be related to the results from other comparable research proceeding simultaneously but not labelled as IBP projects, so that syntheses could include assessments of the 'state of the art'. To achieve a synthesis in this comprehensive form, even for a small part of the total programme, involves extensive work, preferably whole-time for months or even years at a stretch, and in a few exceptional cases this has in fact been done and will be reflected in the subsequent publications. However, in other cases synthesis had, of necessity, to take a different form and will consist of a series of essays on the subject matter of the programme, referring to IBP as well as other results. Between these two approaches there will be a number of inter-mediates, for no attempt has been made at complete uniformity. The problems of preparation and presentation common to the whole series were discussed extensively and intensively, but the way synthesis has been achieved in each case has been largely at the discretion of section conveners and the volume editors, each of whom was appointed by SCIBP.

In any case synthesis is based on data, and during the process of synthesis it has become apparent that a considerable proportion of the data collected during IBP does not lend itself to synthesis. Much of the data collected by ecologists in the early days of IBP was necessary for an understanding of national projects and was not necessarily related to any idea of synthesis; but, as the programme developed, new hypotheses were formed and they needed data of a different type. The data previously collected were not wasted, but became a step in the evolution of the programme.

The process of synthesis clarified three lessons for the future: first, the objectives were perhaps inadequately defined in advance, so that some of the data proved irrelevant to the conclusions; second, while much attention was devoted in the early stages to methods of collecting data, not enough was devoted to methods of processing them; third, the effort which is required in synthesis proved to be greater than some parts of the programme could sustain, so that some data which might have contributed could not be fully explored. But in spite of these defects synthesis has been and continues to be pursued with vigour at both the national and international levels, and the publications which result are so important in relation to the whole process of IBP that they will be described here.

At the national level many volumes of synthesis are planned and some already published. It is too early to give details but it may be noted that the total number of volumes sponsored by national IBP committees, now either published, or in press or planned, exceeds a hundred by a considerable margin. The countries most active in this are USSR, USA, Japan, Canada, UK and Scandinavia. National syntheses are listed in Appendix 5c wherever information on their preparation has been supplied.

The international series

At the international level, everything focusses on the series of volumes which are now in process of publication by the Cambridge University Press. This series, which is expected to run to nearly forty volumes, is being produced with as little delay as possible, but there are some difficulties. Thus, whereas the last two years of IBP until July 1974 were planned for synthesis, in some countries the completion of operational projects extended into these years, and even after them, with consequent delays in synthesis. Moreover, five major IBP symposia of results were arranged after the formal ending of the programme, during September 1974 at The Hague in association with the first International Congress of Ecology. Nevertheless, the majority of volumes in the series are, at the time of publication of this first group, either already with the Press or in an advanced state of preparation. The Special Committee for the IBP, before its demise in July 1974, recommended April 1975 as the last date for the completion of typescripts.

To publish such a large amount of information and discussion so soon after the end of a ten-year programme of research involving some thousands of workers, obviously has disadvantages. The biggest of these is that synthesis has to be completed and written up before a substantial part of the research on which it is based has reached its primary publication stage. This has meant that a number of important projects in the programme have not been included in the synthesis or have been mentioned with little more than title. However, the arguments for early publication were strong and proved to be overriding. The strongest reason was that other major international programmes in environmental biology are already in the planning stage, if not actually started, and it is of great importance that all the scientists concerned in them have at least an outline of the results and conclusions from IBP available for study. This is of particular importance in relation to the Man and Biosphere Programme of which UNESCO is the prime organiser. As an intergovernmental programme, in contrast to the non-governmental IBP, MAB is likely to involve many scientists who have not been a part of IBP, although many of the projects within MAB in fact had their origins in IBP.

Another important consideration was that the world network of biologists which has come into being through IBP is deeply involved in the synthesis process. Parts of this network may be dispersed or at least regrouped in connection with other activities fairly soon; so that synthesis to be effective should be done while the network is still in being. Indeed, a major difficulty is the drift of key IBP scientists to other activities before they have opportunity to complete their share of synthesis. Only in rare cases have scientists been employed to do international synthesis which has, for that reason, often had to take second place to other demands. Then there was the financial consideration that a synthesis on this scale inevitably involves a large number of consultations and small editorial meetings and this costs money. It was best therefore to get it done while the organisation and financial system of SCIBP, however modest, were still operative.

Synthesis does not cover all of the IBP, because some of its studies were somewhat separate from the rest and therefore not amenable to this process. It does, however, attempt to cover all those subjects which were studied in a number of places and for that reason have results to contribute derived from comparing different situations. One principle is that these volumes should not repeat primary data although such data need to be summarised. For example, ecological descriptions of particular sites will become available in research papers, and so, where necessary for the purpose of synthesis, will be in abbreviated form. Another principle is that the presentation is intended to make the volumes readable by the general biologist and others who are interested in the environmental and human sciences. It is not intended that specialised knowledge of any particular subject is necessary for an understanding of the synthesis.

A principle aimed at is uniformity in the use of terms, symbols and units of measurement, following where possible the International System of Units (SI). Such uniformity, though applied with reasonable success in the physical sciences, is much more difficult in biology, especially production biology and ecology, and several discussions which examined the subject at an early stage in the programme, found it difficult to reach any conclusions at all. However, during the synthesis phase a working group was set up by SCIBP and tackled the subject with great diligence and thoroughness. They produced a report which has been widely circulated outside and inside the IBP in several drafts and more recently in the form of a preprint. This is now published as Appendix 6 to this volume. Not every biologist will agree with every detail in that Appendix and not every volume in the series follows it precisely; but where there has been reason to use other terms and symbols, they are accompanied by their own definitions. It is hoped that the report will prove a useful guide now and in the future.

Each volume in this series has a chief editor and often there are one or more co-editors in addition. In cases where a single chapter incorporates the contributions from a number of scientists, there may have been a chapter editor necessary as well. All contributors are named in each volume.

The volumes currently planned for the series are listed in Appendix 5*d* with their chief editors. Each of them has emerged from the work of one or other of the seven sections into which the programme was divided, and as such, the International Convener of each section is basically responsible. However, in this process, SCIBP tried to bring the seven sections together rather than to perpetuate their separateness, and it is possible that the list may ultimately have the addition of a few volumes which are essentially intersectional in character, as already are those volumes which deal with the processes of productivity. For this same reason, the order of listing the volumes does not tally precisely with the sectional order, for it is better to follow an 'ecological' sequence. The number attached to each volume will not, however, coincide with the listing, for all volumes will be numbered in accordance with the order of publication.

An 'ecological' sequence is not easy because most of the subject matter is evolving so rapidly that groupings change. Even obvious borderlines, as between land and water, are arbitrary and, during IBP, efforts were made in some countries to get away from that division and to consider land and water as interrelated parts of catchment areas rather than as terrestrial ecosystems on the one hand and aquatic on the other. Similar considerations apply to some coastal lands and inshore seas or estuaries where interdependence of biological or human processes is a dominant factor.

After this introductory volume the list commences with the major terrestrial biomes, namely Woodlands, Grasslands, Tundra and related habitats, and Arid lands. In some respects these formed the real core of the IBP and the synthesis of work on each biome will in several cases run to more than one volume. Then come several volumes on groups of organisms which were particularly studied in the IBP, namely Small mammals, Granivorous birds, some of the more important Social insects, and it is convenient to add here a separate volume on the Biological control of insect pests. The volume on Wetlands, a term which is used here for shallow-water areas with emergent vegetation as well as swamps and marshes, is a convenient link between the preceding volumes on terrestrial systems and the following ones on aquatic. Freshwater ecosystems are the subject of a particularly thorough synthesis contrived in a single large volume, while those on Marine production mechanisms and Man's influence on the marine environment will be shorter and separate. They are followed by two groups of marine organisms which have been prominent in some IBP studies and offer particular opportunities for development as food supplies in developing countries, namely Mussels and Mullets. These two, like the volumes on particular groups of animals which have come from the work of Section PT, may appear somewhat specialised; but it has proved advantageous to examine the problems of production biology from the grass-roots, so to speak, with a group or a single species at the centre of the theme, as well as taking the overall view of the ecosystem.

Then we have a group of volumes on the processes of biological productivity which apply right across the board, namely Photosynthesis and productivity in different environments, Nitrogen fixation by free-living micro-organisms, Symbiotic nitrogen fixation in plants, and Decomposition and soil processes. Conservation has been prominent in IBP thinking from the very beginning, and a volume of synthesis which is emerging from Section CT focusses particular attention on the scientific basis of Conservation of ecosystems, which has been a major theme of the IBP, rather than the conservation of particular species which is already well looked after by IUCN's Survival Service. The conservation of genes takes the problem several steps further than that of species, and was a prominent subject in the IBP, hence the volume on Crop genetic resources. It is important to note in addition that the conservation of nature in all its forms also entered into other branches of the programme.

Food protein sources helps to link together the two objectives of the IBP as a whole, namely biological productivity and human welfare. The series continues with the five volumes on Human biology of circumpolar peoples, High-altitude peoples, Human growth, Physiological function, and finally, Population structure.

In nearly every case, the volume of synthesis has emerged from an international group which had been appointed during the progress of IBP by the section committee concerned. In each subject, when initiating the synthesis, meetings have been held to assess results. Sometimes such meetings have taken the form of symposia of a hundred or more scientists; at other times they have consisted of small groups of persons who were likely themselves to take a prominent part in the synthesis operation. These groups have drawn up synopses, suggested main contributors to particular chapters, and recommended the overall editor or editors of the volume concerned. If approved by the section convener, the names of the overall editor(s) go forward with the synopsis for formal appointment by SCIBP via its Editorial Committee. Thus the internationality of the series has been assured and the selection of the editors for and contributors to each volume has been based on competence, ability and willingness.

Transfer

Some projects undertaken within IBP were specifically designed and planned as short-term – the data to be collected, experiments conducted, the results analysed and written up together with conclusions drawn, all within a brief span of years. It was, however, recognised from the beginning that the stimulus which the programme provided would initiate some research that could not possibly be completed within a few years and for which appropriate sponsoring organisations would have to be found after IBP came to an end. In the event the projects and activities did, in fact, fall into these two functional groups, but those which must obviously carry on are more numerous than expected, for the world of today is far more conscious of the environment than the world of ten years ago.

The process of transfer from IBP to other organisations was initiated in the early years of the programme and has been going on since then, because from time to time it became apparent that agencies other than SCIBP would be in a position to advance a particular subject more rapidly. One or two examples may be given. One of the first to be transferred was the interdisciplinary subject of man-made lakes, their effects on the environment and on mankind. In this subject hydrology, engineering, geography, geophysics, biology, medicine, agriculture, fisheries, forestry, and sociology are all involved and several scientists who had been concerned in planning studies to be undertaken on large man-made lakes in the tropics under the aegis of the United Nations Development Programme, were also deeply concerned with ICSU. At about the time that IBP started ICSU decided to take up the study of large man-made lakes and, after consultations which revealed that the physical aspects were already reasonably well known but the biological effects were definitely not, asked IBP to develop the subject. A first international symposium on large man-made lakes, designed to open up the problems, was organised by IBP personnel before there was a system for holding specificially IBP meetings, so that the Institute of Biology in London kindly arranged it and published the volume of proceedings (Lowe-McConnell, 1966). A few years later the IBP National Committee for Ghana, in consultation with other national and international bodies, arranged a second symposium on the subject in Accra, and this had a rather particular focus on the problems as they affected countries in Africa (Obeng, 1969). At this point ICSU's Committee on Water Research (COWAR) showed a particular interest, and so IBP handed the subject to them and this resulted in the third and largest international symposium, held in 1971 at Tennessee (Ackermann, White & Worthington, 1974). On that occasion there were many and varied contributions which warranted a more thorough analysis than was possible during a week's meeting, so that COWAR handed the growing child to ICSU's Committee on Problems of the Environment (SCOPE) which had recently been formed and was seeking a few important subjects for initial concentration. The result was a very useful report which highlighted the environmental and human problems concerning large man-made lakes (SCOPE, 1972). Thereafter SCOPE handed the child, now approaching maturity, back to COWAR, which is currently pursuing certain of the problems in a more thorough way then had been possible before.

Another example of early transfer of a subject from IBP was concerned with proposals to monitor changes in the environment. Ideas on this had been forming during the early years of IBP and were first discussed in some detail at the Third General Assembly held in 1968 at Varna in Bulgaria, on the initiative particularly of Scandinavian scientists. It was suggested that extensive activity by IBP on monitoring changes would not only have intrinsic importance itself but would also have a cohesive influence on other diverse projects. Human influences on the natural environment are great and continue to increase rapidly. Many influences are to the bad, such as pollution in its various forms and accumulation of toxic chemicals, destruction of protective vegetation; but other changes may be to the good, such as landscape planning of derelict land, anti-erosion measures, and, when carried out with perspicacity, major engineering works. In some cases the physical components of such changes were already being monitored, at meteorological or geophysical stations; but the biological components, which are often of greater importance, were not. It was clearly desirable to measure such changes against known ecological standards and to make collections of biological data and material for subsequent back-reference. Baseline recording stations at sites chosen with great care might become focal points for ancillary ecological work leading to a better understanding of the biota which make up the environment of man, as well as the changes to which they are subject.

While these ideas originated in IBP, it was quickly recognised that the task would be of long duration and would involve expenditure at a level only possible through governmental and intergovernmental agencies. Accordingly, this project was handed for close examination to the newly formed SCOPE which produced an initial report (SCOPE 1971), and this in turn became an important document for the UN Conference on the Human Environment in 1972 at Stockholm, which was the first occasion for major intergovernmental discussion on monitoring environmental change. Subsequently, with the creation of UNEP, SCOPE produced a further detailed report and, following the intergovernmental conference held in February 1974 in Nairobi, the subject became a major item in UNEP's international programme.

A direct form of transfer, in this case of an active operational programme, was that of CT's world check sheet survey which was handed over as a going concern to IUCN in 1972 at the end of Phase II of IBP, together with its data bank which had been funded and staffed by the Natural Environment Research Council of the UK. In a different category was an incipient journal, the *Marine Pollution Bulletin* which started as a multigraphed newsletter issued by an active PM worker, R. B. Clark of Newcastle University, with the aid of a small IBP subsidy, but was soon handed over to Macmillan Journals Ltd, the publishers of *Nature*, and now, as a regular monthly, has a wide circulation. Other examples of transfer in PM include the handing over of marine conservation from the embryonic Project Aqua Marina to IUCN and it was also agreed that work on marine mammals could be passed to the working party of ACMRR under Sidney Holt.

Table 1. *A provisional plan of transfer of IBP interests to permanent organisations*

Subject	Non-governmental organisation (NGO's)	Intergovernmental organisations (IGO's)*
Woodlands	SCOPE working group on ecosystem studies IUBS/INTECOL	UNESCO/MAB Projects 1 and 2 FAO/Forestry and Forest Industries Div.
Grasslands	SCOPE working group on ecosystem studies IUBS/INTECOL	UNESCO/MAB Project 3 FAO/Plant Production and Protection Div.
Tundra and related habitats	SCOPE working group on ecosystem studies IUBS/INTECOL	UNESCO/MAB Project 6
Arid lands	SCOPE working group on ecosystem studies and plans for concentrated study on arid lands COWAR project on irrigation of arid lands in developing countries IUBS/INTECOL	UNESCO/MAB Project 4 FAO/Agricultural Department, several divisions
Small mammals	A proposed working group within IUBS	Contributes to UNESCO/MAB Projects 1, 2, 3 and 6
Granivorous birds	A proposed working group within IUBS	Contributes to UNESCO/MAB Projects 1, 2 and 3
Social insects	IUBS/IUSSI ISSS, especially for termites	FAO/Plant Production and Protection Div. Contribution to some UNESCO/MAB Projects
Ecology of wetlands	Of interest to IUBS/SIL, ISSS, conservation aspects to IUCN wetlands working group	Contributes direct to UNESCO/MAB Project 5 Links with IHP.
Freshwater ecosystems	IUBS/SIL Conservation aspects (Project Aqua) to IUCN wetlands working group	Contributes to UNESCO/MAB Project 5 Links with IHP
Photosynthetic systems	IUBS/IAPP	Contributes to UNESCO/MAB Projects 1, 2, 3, 4, 5 and 6
Nitrogen fixation	IUBS/IAPP SCOPE working group on biogeochemical cycling IUBS/World Federation for Culture collections (Intra Union Commission)	Contributes to UNESCO/MAB Projects 1, 2, 3, 4, 5 and 6, also Project 8 (World Catalogue of Rhizobium Collections) Co-operation with FAO/Plant Production and Protection Division

Topic	Organisations	Contributions / Links
Decomposition and soil processes	SCOPE working group on biogeochemical cycling	Contributes directly on UNESCO/MAB Projects 1, 2, 3, 4, 5 and 6
Conservation of ecosystems	Servicing of CT data bank and check sheet survey transferred to IUCN	Contributes to UNESCO/MAB Project 8
Crop genetic resources	Contacts with IUCN and IUBS Int. Assoc. Botanical Gardens	FAO/Plant Production and Protection Division; UNESCO/MAB Project 8; Links with UNEP
Biological control of insect pests	IUBS/IOBC	Contributes to UNESCO/MAB Project 9; FAO/Plant Production and Protection Div.
Food protein sources	IUBS/Division of Microbiology	FAO/appropriate divisions
Aerobiology	IUBS/IAA	Appropriate divisions of WHO and FAO
Marine production mechanisms	IUBS/IABO; SCOR, especially Working Groups 24 (Est. of Primary Prod.) and 29 (Monitoring life in the Ocean)	UNESCO Division of Oceanography and IOC (especially the International Decade of Oceanographic exploration); Contributes to FAO/Advisory Committee on Marine Resources Research (ACMRR)
Man's influence on marine environment	SCOR, especially Working Groups 39 (Pollution in Marine Env.) and 45 (Marine Pollution Research); IUBS/IABO; Links with SCOR	UNESCO Division of Oceanography and IOC/IDOE, FAO/ACMRR
Cultivation of marine organisms (mussels and mullets)	IUBS/IAHB; SCOPE working groups	FAO/Fisheries Department, appropriate divisions FAO/ACMRR
Human biology, including: circumpolar and high-altitude peoples, human growth and physique, human physiological function, population structure		WHO, appropriate divisions; UNESCO/MAB Projects, especially 6, 7 and 12

* In addition to the listed organisations, are contacts with UN World Conferences on Population, Human Settlement and Food Supply, with UNEP, WMO and other branches of UN.

Much of the process of transfer consisted of handing on the spirit of IBP which had been engendered by the network of scientists; and the method of doing this was to identify ongoing organisations which already existed for many of the subjects and arrange for them to take a continuing interest in the ideas and activities which had been engendered by IBP. In cases where no such organisation existed, IBP scientists sometimes took the initiative in creating one and then getting it recognised by, or affiliated to, the appropriate ICSU union.

A topic, or rather a group of topics which was a centrepiece of IBP, and of which the ideas were transferred at a comparatively early stage and are now being developed extensively by other organisations, relates to the interdisciplinary study of the major biomes which comprise the biosphere and the comprehensive analysis and synthesis of the ecosystems of which they are composed. IBP's own work in these matters is referred to in Chapter 4 and elsewhere; but the transfer of the ideas started in September 1968 at the intergovernmental conference convened by UNESCO in Paris on the 'Scientific basis for rational use and conservation of resources of the biosphere', a conference which led subsequently to the MAB programme which adopted and developed many of these ideas.

MAB is an intergovernmental programme, but it aims to invoke the assistance of non-governmental as well as governmental scientists. In the planning stages each of the thirteen main projects of MAB passes through the process of a panel meeting which elaborates the scientific ideas, and then through a working group which consists of nominated national participants who are in a position to indicate the special views and interests of national MAB committees and later perhaps to initiate actual research. The process of IBP transfer to MAB therefore ensured that IBP experience and expertise in each of the main projects was fully available to the MAB panels and working groups. To this end a good many memoranda were prepared for MAB by IBP scientists, some of whom also participated in the various meetings. A number of these transfer activities were funded through short-term contracts between UNESCO and IBP.

Similarly close contacts were established between specialist agencies of United Nations other than UNESCO, and these may become the residuary legatees, so to speak, of certain branches of IBP. The examples of FAO and the IBP work on plant gene pools is recounted in Chapter 4 (pages 86–88), and there are close relations also between some branches of WHO and IBP's studies on human biology.

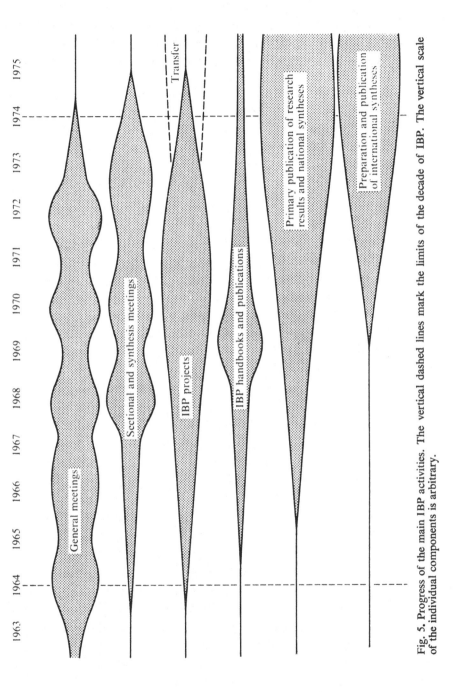

Fig. 5. Progress of the main IBP activities. The vertical dashed lines mark the limits of the decade of IBP. The vertical scale of the individual components is arbitrary.

A list of the main fields of the IBP is related in Table 1 to the non-governmental and intergovernmental organisations which are or may be in a position to sponsor follow-up activities. The progress of IBP activities towards synthesis and transfer is illustrated in Fig. 5.

6. Publications

A broadly based international programme like the IBP would clearly result in a large bulk of publications, and how to arrange for them in such a way as to be of maximum use during the programme itself and in subsequent years, offered several different choices. One method would have been to start a series of volumes as a channel for publication of all IBP scientific results, a series which would run not only for the duration of the programme but for as long afterwards as the results were forthcoming. This method was considered during the early stages but was discarded on three grounds: it would have involved heavy finance which was not in sight; it would have been difficult to devise an adequate system for vetting the submitted contributions, such as was provided by scientific journals, in order to ensure a high scientific standard; and the experience of IGY showed that such a series would have needed to run for many years after the end of the programme. Therefore we looked at an alternative method, namely for SCIBP itself to publish almost nothing, but to rely on the regular continuing service provided by scientific and technical journals and commercial publishing houses. This would have been an easy way out, but would hardly have met the anticipated needs. Accordingly, an intermediate system was adopted whereby IBP would make full use of all existing publication facilities, especially for the primary publication of research results, but would also run limited series of its own to meet its own particular needs.

The result of this policy was that the publications resulting from the IBP fall into six main categories as described below:

SCIBP publications

IBP News

IBP News was published direct from the IBP Central Office in London, and was the official periodic organ of the programme. It was started in 1964 to spread information around the world about the content of the programme and how it was being organised, and has continued throughout the decade with two or three numbers each year, making a total of twenty-five. An appropriate number of copies of each part was issued free to participating countries and to key IBP scientists, with other copies on sale. Most numbers contain the records of major IBP meetings,

117

sometimes in summary form, and other information which it was essential to spread around participants. Numbers of particular importance are no. 2 and no. 9 which contain the plans for Phase I and Phase II of the programme in its preliminary and revised form, and no. 24 which contains the plan for Phase III.

Newsletters

The Biosphere and newsletters served for the quick dissemination of information about the IBP. *The Biosphere* was published from the Central Office as a newsletter in English and French as the two official languages of ICSU, but later reduced to English only. It was started in 1967 and ten numbers were published during the next five years; it was widely distributed free of charge, and correspondence indicates that it was much appreciated. However, in 1971 SCIBP was passing through a particularly lean phase in which the income by no means met all the demands for scientific co-operation, and coupled with this the load of work on the very small international staff meant that something had to go. In consequence, further publication of *The Biosphere* was postponed, and in spite of requests for its resuscitation, this did not prove feasible.

In various branches of the programme a number of other newsletters were issued at more or less regular intervals, some printed and other in multigraphed form. Several of the international sections kept scientists generally informed of developments by such means, and some participating countries issued IBP newsletters themselves. Among the latter, for example, was *Inter-American IBP News* which, though published from the USA, was concerned with the New World as a whole and served a good purpose in South America as well as North. Other notable newsletters which served regional in addition to national functions were issued by Sweden and by the USA, especially from its HA section.

IBP Review

In August 1968 at the time of UNESCO's Biosphere Conference, which resulted some years later in the initiation of the MAB programme, a booklet was published by SCIBP with the primary purpose of informing that conference what IBP was about. It proved popular and became known as the *IBP 1968 Review* as the beginning of a series of summarised review booklets which were subsequently published in 1969, 1970 and 1972. There was no need for a subsequent issue since most of the

118

information which the *IBP Review* presented is contained in more extended form in this volume.

Handbooks

The series of IBP handbooks, which numbered two dozen by the end of the programme (Appendix 5*a*), was designed for, and served, a quite different purpose. It was started in response to the need for guidance in many parts of the world on appropriate methods of research on IBP subjects of a kind which would give results which were comparable one with another. To begin with there was some tendency to focus on rather unsophisticated methods which could be used in remote places as well as in well equipped laboratories, and which could be described in rather small inexpensive books. But as the programme evolved and in some subjects new or improved methods of research and analysis of results became available, the handbooks tended to expand in their individual size, to include sophisticated methods as well as simple ones, and in consequence to become more expensive to purchase. Thus the changes to which terrestrial ecology has been subject during the programme is to some extent illustrated by the IBP handbooks. Early volumes produced by Section PT, advising on methods for studying biological productivity in woodlands and grasslands, appear now to be rather simple stuff. But in the early stages of IBP they were much welcomed. At that time the big biome projects using the techniques of systems analysis and mathematical modelling were hardly thought of, and indeed, the collection of quantitative ecological data which made such techniques possible was not at all common. If a new series of handbooks on methods for studying productivity of terrestrial ecosystems were now to be produced, they would look rather different, and this shows the rapidity with which terrestrial ecology has changed during the period of IBP.

On the other hand, the IBP handbooks which emanated from Section PF appear in a rather different light. They deal comprehensively with methods of inland water research covering chemical analysis of waters, primary production, secondary production, fish production, microbial decomposition (IBP handbooks nos. 3, 8, 12, 17, 23) together with *Project Aqua: a Source Book of Inland Waters Proposed for Conservation* (no. 21). These books have been widely acclaimed by limnologists and have run through reprinted or revised editions. If newly written today they would probably not appear very different, except perhaps in the rapidly evolving subject of microbial ecology, and this perhaps illustrates that aquatic ecology, as a discipline, was more fully formed than terrestrial ecology at the start of IBP.

A few handbooks took the form of general explanation about sections of the programme and how the work was arranged and in consequence are of more ephemeral value. However, the large majority dealing with methodology are likely to continue, in reprinted, revised or rewritten editions, for a considerable time to come. Some of them are definite landmarks in the subjects about which they are concerned and some have been adopted as texts for academic teaching and as such have achieved commercial success.

Until near the end of IBP this series of books was financed by SCIBP and published on SCIBP's behalf by Blackwell Scientific Publications Ltd of Oxford, who arranged the printing, distribution, advertising and sales. All the editors and contributors gave of their services without financial recompense, as personal contributions to the success of IBP. At the formal ending of the programme in 1974, the system of publication was changed in order to ensure continuance of the series as long as it remains useful. Blackwell Scientific Publications Ltd accepted financial responsibility for Handbook no. 24 and for new editions of existing numbers and perhaps new books to add to the series, and they pay appropriate royalties which are divided between ICSU and those biologists who do work on the books.

Primary publication of IBP results

In terms of achievement as a whole this category of publications is the most important and, once the decision was taken *not* to publish an independent series of IBP results, it has been left entirely to the discretion of national authorities, including the IBP committees, institutes and individual scientists, who have been responsible for conducting the research work. In consequence there is considerable variety in the form of publication. Some of the major biome research units, particularly those in Canada and the USA, have issued a very large mass of reports prepared by hundreds of research workers and containing large quantities of primary data as well as descriptions of the research and the conclusions derived therefrom. Many of these reports are in multigraphs form with limited circulation. In due course doubtless they will be added to or replaced by papers published in appropriate scientific and technical journals or in books. In the USSR a series of 50 volumes have already been published, covering all sections of the IBP and many of them include the primary publication of results as well as in many cases a synthesis on the national basis. They are in Russian language but generally with the contents sheet or summaries translated into English.

Most countries however, have relied on the regular system of scientific societies and journals to publish the scientific results of IBP in the form of papers submitted to them, no different from any other research papers. Since there are more than two thousand IBP research projects listed in *IBP News*, nos. 13–20 and 22 and considerably more than half of these have become truly viable and are producing scientific papers, sometimes many from a single project, the total number of papers already runs into some thousands and is likely at least to double during the next few years. They are very widely scattered in the scientific press, published in many languages, and therefore difficult to keep track of although efforts have been made to collect them comprehensively at the Sectional and Central Offices of IBP. One set, as complete as possible, is for permanent reference, housed at the Linnean Society of London at Burlington House.

Although no scientific journals were started by IBP itself, in several fields the need to publish results helped to start and successfully to develop new international journals. An example is *Photosynthetica*, published from Prague since 1967, in which a considerable proportion of IBP's photosynthesis papers are published. This has provided a new international forum in this field of research.

Proceedings of symposia and meetings

The number of scientific and technical meetings, including symposia, undertaken under the auspices of SCIBP has reached 170 during the decade (see Appendix 4). Most of these were convened for specific purposes and many led to publications such as those contained in the handbook series and the international synthesis series, or, in the case of those associated with IBP assemblies and meetings of SCIBP, they were published in *IBP News*. However, there were many international symposia of which the purpose was the discussion of plans and progress, or of IBP results, and in many cases these symposia were deemed worthy of publication as volumes in their own right. At the national level a considerable number of symposia of IBP results have been published, for example several convened by the Royal Society of London and published in the Royal Society's *Transactions*.

SCIBP, with already too many calls on its resources, has only in rare cases accepted any responsibility for the publication of IBP symposia on results, and those which have been published have been financed and organised on a national basis, generally by the country which was host to the symposium in question. In certain countries, such as Poland, the publication and distribution of such volumes has been a major contribution to the international work of IBP.

National programmes, reports and synthesis

This category of publications is of high importance, but so unequal among participating countries that the list (Appendix 5c) is no more than a rough guide to intentions in the form of programmes, or achievements in the form of reports and synthesis. The reason is of course, that different countries have different ideas on what should be published and also different financial resources. However, bearing in mind that the number of entries under each heading is not necessarily correlated with that country's activity in the programme, it does give some idea of the coverage. The list is restricted to publications of a general nature and does not include the very large numbers of sectional or project reports which have come from some countries.

International synthesis

And so we come to the series of international syntheses published by the Cambridge University Press of which this is Volume 1, and of which the purpose and content has already been described in Chapter 5.

Public relations

The above five categories are concerned with the scientific publications of the IBP, but in addition, there has been a considerable number of publications designed for the interested public who have general rather than special interest in the subject matter. These have included reports of lectures, discussions, broadcasts and television talks, articles intended to publicise certain aspects of the programme, and a few documentary films. The public relations aspect of the programme is not regarded as one of its successes, but rather as a disappointment. Almost throughout it has been carried on by scientists who have been fully occupied with other responsibilities but who managed to find time for publicity. Only for short periods have professional science writers or journalists been associated with IBP, mainly because the funds did not suffice. For a time the monthly journal *Science Journal,* published from London, ran articles on IBP topics, selecting those which the editor considered to have a wide interest. This was followed later by a similar arrangement with the American-based *Bioscience.* One or two national committees have been in a position to employ staff who have had special interests in public relations. In general, however, the public relations of IBP have been among its weaker points. There has been overmuch reliance on the inherent interest of the subject matter which it had been thought would provide a linkage with the media. With hindsight it may be said that a major international programme of this kind, which depends to a considerable extent on public and government support, should give more attention to projecting its image. Now that the programme is ended, it is noteworthy that several participating countries, in addition to SCIBP itself, have plans for presenting the results in books intended for the general reader in addition to the scientific syntheses.

7. Financing the programme

Ronald Keay

At a superficial level the situation has been all too easy to describe: 'never anywhere near enough central funds to support the plans of the scientific leaders!' Looking back, one may well feel how mean and short-sighted were the various fund-granting bodies, but it is easy to forget how novel the idea of IBP was when it first emerged in the early 1960s. Never before had there been an international and interdisciplinary collaborative research programme of this kind in biology. The relevant Unions were impecunious and attempts to form an international organisation for ecology – a subject in which international co-operation would seem particularly important – had been sadly ineffective.

The financing of IBP was first discussed at the 13th Executive Board of ICSU in September 1961 at the same time as the agreement to go ahead with the programme. An estimated budget of $20 000 for 1962 was proposed and accepted on the basis that $15 000 would be provided as an ICSU loan and $5000 as an ICSU grant; no plans for long-term funding of the programme were enunciated. Further discussion by correspondence followed the first meeting of the IBP Planning Committee in May 1962 and at one stage it was proposed that a limit of $3000 to $4000 be placed on expenditure by each of the sections from 1962 to 1964. By early 1963, the conveners of the sections felt strongly that the lack of funds was severely hampering the development of a viable programme. In particular, it was felt that the restriction on funds precluded the establishment of sufficiently well elaborated programmes of activities to interest Foundations. An urgent financial need was for the conveners to have assistance from full-time IBP staff known as scientific co-ordinators.

In spite of these difficulties the convenors succeeded in making proposals for the various sections of IBP and these were discussed at the first General Assembly of IBP in Paris in July 1964. This meeting passed the resolution quoted on page 141 which had been put forward by C. H. Waddington, then President of the International Union of Biological Sciences. The purport of that resolution was that, during the first two preparatory years a minimum of $250 000 to $500 000 per annum would be needed to cover the international expenses which were of course additional to funds required in each participating country to establish

125

national committees and to initiate research. In fact, the central income of SCIBP only once exceeded $250 000 a year and that was in 1970, well into the operational phase of the programme, and the preparatory phase (1964–7) had to manage with much less. It became apparent then, and throughout the programme, that the international activities depended as much on local support from national and other sources as on the funds which came directly to SCIBP and were administered by the central and sectional offices.

Such national support, which did not appear in the centrally audited accounts, included:

local support for international scientific meetings, including organisational expenses, living expenses for participants, publication costs, etc.;

support for the travelling expenses of scientists attending international meetings;

provision of modest office accommodation and secretarial support for conveners and others with international tasks to perform;

support for scientific work in the synthesis phase of IBP.

The list of countries providing such support is a long one and it is perhaps invidious to mention any. However, provision of accommodation for the Central Office by the Royal Society of London and of support for the co-ordination of the PP section by the Czechoslovak Academy of Sciences, and of CT office and international data centre by the Nature Conservancy of the UK, may be mentioned as examples. This support is difficult to estimate in financial terms, but in several cases it included provision of staff and complex equipment as well as accommodation, so it was certainly substantial. It was particularly important at the beginning of IBP.

The main sources of support for the international co-ordination of the programme may be seen in the following summary of income to the central account in the years 1962 to 1974, added together.

	US $	*Proportion* (%)
National dues and special contributions	822 342	43.70
Grants and loans from ICSU	283 780	15.08
Contracts with UNESCO	299 564	15.92
Publications	231 913	12.33
Other income	244 070	12.97
Total income 1962–74	$1 881 669	100.00%
(1962–73 audited, 1974 estimated)		

The above summary shows that by far the largest part of the central income came from national support through adhering academies and research councils. In all, fifty countries paid dues in categories ranging from $100 to $25 000. Starting with $6000 in 1964, the annual income from national dues rose to a maximum of $115 567 in 1970, a figure substantially more than that received by any other committee or union of ICSU (except the International Union of Pure and Applied Chemistry). In addition to paying dues in the highest category, the US National Academy of Sciences made special contributions of $50 000 a year for the four years 1965–8; this generous support was particularly valuable in establishing the central and sectional offices.

Support from ICSU was in the form of grants, but also in the early years as interest-free loans. A total of $67 750 was loaned by ICSU in the years 1962–6, and repayment by SCIBP at a nominal rate began in 1967. By the end of 1974 a total of $35 000 had been repaid and there is every expectation that the assets of SCIBP and the royalties from its publications will more than cover the outstanding sum. The funds that ICSU loaned to SCIBP came mainly from grants received by ICSU from the Ford Foundation and the Nuffield Foundation and UNESCO, specifically to support such scientific programmes as IBP. Although the loans sometimes seemed like a weight tied around SCIBP's neck, it was appropriate that at the end of the programme the assets and royalties should return to ICSU and so become available for the support of new programmes in the future. Having regard to the slender resources of ICSU, the financial support it gave to IBP, especially in the years 1969–71, was generous and invaluable.

Direct contracts between SCIBP and UNESCO, for the support of meetings and other projects of mutual interest commenced in 1965 and amounted to as much as $56 500 in 1970. Individual contracts seldom amounted to more than $5000 and were negotiated each year. This tangible evidence of UNESCO's growing interest in the programme developed into close co-operation between IBP and the UNESCO Man and Biosphere Programme which was particularly valuable during the Phase III process of synthesis and transfer.

In the early years of IBP, publications were a major element in the expenditure budget. Financial arrangements for the IBP handbook series, published by Blackwell Scientific Publications Ltd, entailed SCIBP paying the full costs of producing each volume, so there was inevitably delay in balancing the account. SCIBP had also to produce, with no prospect of adequate financial return, various other publications, such as *IBP News* and *The Biosphere*, as well as some sectional publications such as reports and some proceedings of meetings. Some of the IBP handbooks were, however, very successful from the financial point of view, and from 1972 onwards the publication programme provided SCIBP with a net income. By 1974 all costs in producing the handbook series were complete, and continuing sales bring a useful profit.

The expenditure side of the SCIBP central accounts for the years 1962 to 1974 may be summarised as follows:

	US $	Proportion (%)
Sectional activities	1 049 262	56.29
Meetings of SCIBP, the Bureau and General Assemblies	118 539	6.36
Central Office (including officers' travel)	434 557	23.31
Publications	226 798	12.17
Repayment of loans to ICSU	35 000	1.87
Total income 1962–74 (1962–73 audited, 1974 estimated)	$1 864 156	100.00%

The officers of SCIBP always endeavoured to devote the greater part of the limited central income to scientific activities, especially the sectional planning and synthesis meetings and the preparation of handbooks. The proportion spent on central meetings and the Central Office was extremely modest; few organisations can have contained their administrative expenses in a highly inflationary period as well as SCIBP. Thus, the amount spent on the Central Office, travel of officers, meetings of SCIBP, the Bureau and General Assembly rose from $56 700 in 1966 to only $68 600 in 1972, after which it declined.

Several efforts were made to obtain support from foundations, but, apart from very modest support from the Ford Foundation and the Nuffield Foundation through ICSU in 1965–6, only the Commonwealth Foundation responded. That Foundation's support from 1970 to 1974, a total of $76 982, was designed to assist scientists from the less developed countries of the British Commonwealth in attending international IBP meetings.

The central accounts referred to above represent only a small proportion of the total cost of the IBP. The main part was incurred by participating countries on their own research projects. Just how much the total expenditure on IBP research amounted to is extremely difficult to estimate, because systems of financing and accounting differ greatly among participating countries, and many of them find it difficult to distinguish their support for IBP projects from other support for science.

Nevertheless, to take one or two examples where costs have been roughly estimated, a conservative figure for the amount spent for the theme concerned with Grasslands, in which twenty-nine countries participated, for a single year of 1970, was about $5 000 000. In the case of the Tundra Biome in which, owing to its geographical distribution, only the nine countries participated, the cost of research over several years was about $1 000 000. For the grand total of the cost of national participation in research, two very rough estimates have been attempted, one from such financial figures as have been made available with appropriate extrapolation, and the other worked out on the basis of man-years. These suggest that during the Operational Phase II, the total was running somewhat in excess of $40 000 000 per annum.

Compared with this, the amount which has been spent on the international co-ordination for the whole programme has been small. As the above tables show, the average annual central account SCIBP budget has amounted to no more than $200 000. If this figure is doubled, to take account of the additional international support which was provided by some of the participating countries, it is apparent that the international administrative costs amounted to no more than about one per cent of the national operational costs, a proportion which can be considered satisfactory.

8. The meaning of IBP for the future

François Bourlière

It is hard to appreciate the real meaning of an ambitious research venture such as the International Biological Programme until its results have been fully published. For the time being thousands of scientists are busy throughout the world collating, comparing and interpreting the data provided by hundreds of research projects. No doubt, these syntheses, carried out at the project level as well as at the national and international levels, will bring out many unexpected results, the implications of which cannot be foreseen now. Furthermore, the IBP decade has coincided with a worldwide interest in environmental problems, their economical and even political implications. Ecology has become fashionable in many circles and an overenthusiastic pedant has even dared to pretend that it has now become the proper study of mankind – a poor paraphrase of Alexander Pope. It is therefore difficult to appreciate the real impact of our programme, and to evaluate within the various promising research trends of the present day's environmental biology, those which were strongly influenced, or even initiated, by IBP contributors.

One way of getting round this difficulty is to look at the current scientific literature and at the programmes of symposia and learned societies' meetings to see what is now the response of the ecological fraternity to the ICSU scientific challenge, ten years after the initiation of our programme. Outside the academic world, it is also easy to note the way IBP has obviously influenced the preparation and the launching of the new governmental, intergovernmental and non-governmental research programmes in the environmental field.

When this is done, it becomes readily apparent that the goal set up a decade ago, the study of biological productivity and its relationships with human welfare, was perhaps somewhat ambitious but has nevertheless proved fruitful. An efficient methodology was proposed and a serious start was taken. In a number of countries this happy development was undoubtedly due to IBP's initiative. The programme helped many a national committee to break through more traditional approaches to ecology, and find the necessary extra money to fund new national research and training.

131

The Evolution of IBP

Biological productivity and IBP

C. H. Waddington has pointed out in Chapter 1 that in the early sixties production biology was not a very popular subject of investigation for the scientific community in many countries. This is true, but the basic concepts were already there, and both in Europe and North America a number of pioneer studies had clearly shown the way such investigations should be carried out (Thienemann, 1931; Lindeman, 1942; Ivlev, 1945; Clarke, 1946; Macfadyen, 1948, 1957; Odum, 1953; to quote but a few). Those who were given the responsibility of building up an IBP research programme on ecosystem productivity had therefore a solid body of knowledge upon which to rely. Their objective was more to refine and make the methodology comparable as far as possible, and organise a comparative study of the structure and functioning of a number of representative terrestrial and aquatic ecosystems located under a wide range of climatic conditions. At the same time K. Petrusewicz and I. Málek strongly supported the view that ecosystem studies should not be limited to the measurement of energy flow and nutrient transfer through the successive trophic levels, but ought also to give particular attention to the population biology of participating organisms and to production processes. This was a very wise proposal, which did not receive enough attention in some national programmes.

The most immediate result of Section PT, PP and PF activities was to provide a conceptual framework and a set of reliable research methods for those, not only ecologists but also resource managers, who need to estimate the potential of biological production of any type of ecosystem, artificial as well as natural, successional (seral) as well as climax. Such a tool is badly needed at a time when large areas of the tropical world, whose ecology remains very poorly known, have to be drastically transformed and exploited on a sustained-yield basis in order to provide future generations with an increased amount of adequate foodstuffs. Needless to say, agronomists, foresters, animal breeders and fishery specialists had already provided their own disciplines with adequate techniques to estimate and monitor some segments of primary or secondary production in some man-modified ecosystems. However, the IBP holistic approach to ecosystem studies was not to consider each of its components in isolation, but as continuously interacting with the other 'compartments' of the system. Precise knowledge of these interactions is particularly important when one is attempting to modify natural ecosystems or to create new agrosystems. Up to now, man's efforts to increase food production have mainly consisted in channelling primary

132

or secondary production of natural systems into human use rather than to the other animal consumers and micro-organisms. This has led to the replacement of complex and well organised communities by simplified ones. Such systems lose the features of the natural and stable ecosystems: the genetic diversity, the many diverse control circuits, the protection against epidemic decimation of their most numerous populations, and the means for soil maintenance and weed control (Odum, 1971). They are consequently more fragile and their functioning and evolution needs to be closely watched. The studies of natural and artificial ecosystems initiated under IBP therefore laid the basis for scientifically grounded 'ecological engineering'. This idea has now been taken over by such intergovernmental programmes as the MAB programme of UNESCO, of which the general objective is to develop the basis for the rational use of the natural resources of the biosphere, and for the improvement of global relationships between man and his environment.

As soon as the various ecosystem projects began to materialise, the need for a means of providing an insight into the inter-relationships within each system, as well as forecasting of change, was very quickly felt within IBP. This led to a renewal of interest for the application of system analysis and modelling to ecology – a trend which is now steadily developing (Watt, 1968; Van Dyne, 1969; Jeffers, 1972). The ultimate goal of ecosystems modelling is not only to understand better their dynamics and the impact of stresses upon them, including stresses generated by man, but also to gain the predictive power necessary for control and optimisation of biological production. If man is going to avoid major ecological catastrophes within the coming decades, he has to learn how to use nature without destroying it (UNESCO, 1972*a*).

At the same time as ecosystem analysis was progressing beyond the mere descriptive stage to reach that of mathematical formulations, population ecology was also reaching a new level of sophistication. A number of US and UK ecologists have been active in trying to provide insight into fundamental problems such as species distribution, community structure and competitive interactions among them. They have developed their concepts in a number of recent and most stimulating books and articles (MacArthur, 1972; Emlen, 1973; Ricklefs, 1973; May, 1973; Maynard Smith, 1974; Cody, 1974). Instead of using computer models of entire ecosystems which, they feel, are so specific that they do not often lead to general laws, they prefer to use mathematical models of particular phenomena. In a second stage they undertake descriptive studies to confirm or extend their models. The contribution of IBP to this other facet of theoretical ecology is not as obvious as to that of ecosystem modelling, but is however far from negligible. The 'horizontal projects' of the PT section (Small mammals, Granivorous birds, Social insects) provided a solid body of supportive evidence for some taxonomic groups of major economic importance.

During the last decade, ecology has thus made a significant step towards appearing as a predictive science and IBP has definitely been instrumental in aiding this significant change, the implication of which for the future of mankind need not be emphasised.

Human biology and IBP

Although less spectacular, IBP's impact upon modern human biology has certainly been as important as that on production ecology. In the early sixties the scientific study of man was still carried out along traditional lines. The physical anthropologists were busy collecting even more data on the morphological characteristics of ill-defined 'racial' groups, the cultural anthropologists were describing the material and social cultures of more or less 'primitive' people (or theorising about them), the human geneticists were somewhat flooded by an expanding number of blood and tissue polymorphisms, and the medical scientists were kept active by the usual public health problems of developing and industrialised countries. But the integrative approach was still lagging behind, despite the efforts of a few pioneers (Sorre, 1943; Dobzhansky, 1962). Very few scientists seriously ventured to study either the influence of some environmental parameters upon certain morphological and physiological characteristics of our species, or conversely to understand the influence of some behavioural characteristics (e.g. mating systems) upon the genetics and dynamics of human populations.

One of the first books to advocate a true holistic approach to human biology was that of Harrison, Weiner, Tanner & Barnicot (1964) whose authors incidentally had already started to play an important role in the initiation and organisation of the HA section. Here at last the major problems facing human biology were clearly stated and this textbook probably did more to convince many people throughout the world of the necessity and feasibility of a fresh outlook than more pretentious papers. The clear and straightforward way in which the Human Adaptability programme was set forth by J. S. Weiner in the first chapter of Baker & Weiner's book (1966) exerted a deep influence well outside IBP circles – all the more since this book provided readers with a wealth of base-line data not previously reviewed in such a comprehensive way. Quite significant in this volume was the emphasis laid upon multi-disciplinary regional studies. This has led to some interesting developments: demographers, social anthropologists, geneticists, physiologists and medical scientists have now begun to work together, as exemplified in the book by Harrison & Boyce (1972) and in the HA volumes of the IBP synthesis series. The relationship between some of the variables studied now becomes clear.

The emphasis laid from the very beginning of IBP upon longitudinal studies of human development was a wise decision. Although only the growth studies were given adequate attention during the duration of the programme, this incentive has now been followed up by those who concentrate upon the whole developmental span of the human life cycle, from birth to old age. A better appreciation of the mechanisms of the so-called degenerative diseases might well emerge from this type of study. It is also to be hoped that the researches upon working capacity and physical fitness, which have up to now remained somewhat at the pilot-project stage, will develop in the coming years. The respective roles of heredity, environmental factors, nutrition, age and training need to be urgently clarified – given the major importance of the problem for developing countries.

The interest shown during IBP for the study of human migration has also had a strong impact upon modern studies of human populations. Not only intranational but international migrations greatly influence gene flow and contribute to the disappearance of isolates in human populations, but they are apparently highly selective processes: the stay-at-home in many ways differs from his migrant relatives. As migration is nowadays becoming a large-scale phenomenon in both developing and developed countries, with many social and economic implications, it is receiving high priority in the new international environmental programmes (UNESCO, 1972*b*).

IBP studies and monitoring programmes

Since the 1972 UN Stockholm Conference the monitoring of environmental parameters is attracting a great deal of attention all over the world. This is amply justified. If mankind is to prevent ecological catastrophes, the decision makers must be regularly provided with the adequate information necessary to ensure the protection of human health and the wise management of the environment and its resources. In August 1971, the intergovernmental working group on Monitoring and Surveillance clearly laid down the objectives for a worldwide environmental monitoring system (UN, 1971): (1) to increase quantitative knowledge of natural and man-made changes in the environment and of the impact of these on man's health and welfare; (2) to increase understanding of the environment, and, in particular, of how dynamic balance is maintained in ecosystems; (3) to provide early warning of significant environmental changes in order that protective measures be considered; and (4) to make it possible to check the effectiveness of established regulatory mechanisms and to plan optimal technological development. Since then a number of expert groups have attempted to propose detailed action plans to achieve these goals.

Unfortunately, most of these plans are restricted to 'priority pollutants' of broad international significance (Munn, 1973), and their assessment in air, water, soil or food. This is utterly insufficient. Without by any means under-estimating the importance of pollution it is hard to equate the few thousand fatal human casualties per year (mostly in industrialised countries) which can be attributed to it with the hundreds of millions of deaths from starvation, malnutrition or disease which, during the same time interval, are the direct consequence of land management practices and policies with no adequate ecological basis. The biome projects of IBP have amply demonstrated that the really meaningful impacts of pollutants upon natural and man-modified ecosystems are those which might lead – through chronic toxic effects, mutagenesis, reduction of species diversity, etc. – to a permanent abatement of the biological productivity of these systems. The difficulty is to identify the target organisms within each ecosystem and for every major class of pollutants, to select the parameters to be monitored and also to find the right way to do it. For the time being, ecotoxicology is unhappily very much in an embryonic stage, and a better knowledge of the structure and functioning of the various terrestrial and aquatic ecosystems, particularly in the tropics, is needed before a scientifically grounded ecosystem monitoring programme can be launched. Such a basic knowledge is expected from some of the MAB projects presently being initiated by a number of countries under UNESCO sponsorship, projects which are largely a continuation and development of some of our most successful IBP pilot programmes. In the meantime WMO, WHO, FAO and other international and national agencies will continue to carry on and improve their monitoring of climatic changes, public health problems, and cereal and fisheries production.

Another IBP initiative which will certainly help some day to organise an efficient network of monitoring stations throughout the world is the CT review and critical evaluation of existing nature reserves and biological stations, now handed over to IUCN. Whatever the physical and biological parameters finally selected to monitor ecosystem changes may be, such monitoring will have to take place in areas whose base-line characteristics are already well known. The IBP survey makes obvious that the existing 'biological observations' are of very unequal value, that they are very unevenly distributed throughout the major biomes, and that new areas and stations will have to be established and adequately staffed. Some of the desirable criteria for site selection and management stand out clearly from this CT exercise.

A testing ground for new environmental research

Whatever the importance for the future of some of the scientific results of IBP may be, they probably do not represent the most significant outcome of our programme. If IBP, directly or indirectly, had such a catalytic influence on the new environmental research programmes now being launched, it is because it initiated a new way of tackling the study of ecosystems and that of human ecology. First of all, and right from the beginning, emphasis was put on interdisciplinarity. Despite the practical difficulties inherent to that type of approach, an effort was made to bring together climatologists, soil scientists, botanists, zoologists and physiologists to draw an action plan for the study of biological productivity, as well as demographers, geneticists, physical anthropologists, social anthropologists and medical scientists to do the same for that of human adaptability. At the same time SCIBP did its best to promote real international participation to our common venture. If we unfortunately failed in arousing interest in many developing countries – largely due to their lack of funds and trained personnel – we were very successful in East–West co-operation. Linguistic or political barriers did not prevent a stimulating exchange of ideas, techniques and results between participating groups during the three developmental phases of IBP. Quite obviously the non-governmental character of ICSU was largely responsible for this happy state of affairs. The scientific fraternity generally enjoys a freedom of thought and action that few governmental organisations can match. This is why all of us feel so strongly that the ICSU family ought to maintain in the future, one way or another, this close and fruitful co-operation between environmental biologists.

137

Another determinant factor in the success of the programme was the happy combination of group planning and personal initiative made possible by the flexibility of our organisation. Once the major objectives were fixed at the SCIBP level, international conveners were appointed who were entirely responsible for the refining of the sectional programmes in close co-operation with the Scientific Director. This was done by successive stages, with the help of various working group meetings and workshops, during which views, experience and results were freely exchanged. Furthermore, these international gatherings benefited greatly from the viewpoints presented by the national committees, some of which were particularly active. Finally, once the sections' programmes were more or less finalised the appointment of a number of full-time scientific co-ordinators made possible a continuous interchange of information between project leaders. The role of these scientific officers during the preparatory, operational and synthesis phases of IBP can never be emphasised enough. Any research programme of some magnitude must nowadays include at its executive level a number of full-time competent scientists, well trained in the corresponding disciplines. It is no longer possible to continue to rely only on the enthusiasm and devotion of part-time bio-politicians or research leaders already overloaded by their national and/or professional duties. And the best conceived programme will very quickly be bogged down in mere paper mess if no competent scientists are able continuously to follow its development.

This is all the more necessary since any active research programme will have to be modified continuously to take into account the developments arising from the programme itself. Our IBP experience again proved that very few of the most successful themes developed as they were originally scheduled.

Last but not least, one obvious cause of the success of some of our projects was the long lasting interest shown, and the continuous support afforded, by UNESCO at the international level, and by some academies and national research councils at the national level. Although a non-governmental venture, IBP enjoyed the active support of governmental agencies in many countries; some more recent intergovernmental programmes would be happy to enjoy similar support now.

Quite obviously we also had our disappointments and failures and these have to be taken into consideration for any future action. Most of them stem from the fact that we were too often trying to do too much in too short a time. The very title of IBP was itself misleading, an International Ecological Programme would have been more realistic, although again too broad. In many cases we also overestimated the extent of our capacity to propose an adequate methodology for the study of some 'compartments' of the ecosystems. In some countries, the national committees contended themselves with putting an IBP label on a number of unrelated so-called 'supporting projects', instead of focussing their efforts upon a few interdisciplinary pilot projects. This should be carefully avoided in any future environmental research programme, all the more since funding problems are now becoming more and more acute.

However, I do not know of any research programme which has ever been 100% successful, and the time factor has always to be taken into consideration. Amongst the handful of seeds sown in the early sixties by those who planned IBP, some immediately sprouted vigorously and these young plants are now producing their first fresh crop of results. Others have undergone a process of delayed germination and are only now beginning to produce their seedlings on new soils. Some cross-fertilisation has also taken place and more will follow in the coming years. This is to be credited to ICSU and our founding fathers. Their enthusiasm and foresight not only initiated an exciting venture which everyone enjoyed, but probably also a new phase in the evolution of environmental biology.

Principal resolutions from General Assemblies and Meetings of SCIBP

From I General Assembly, Paris 1964

Finance for Phase I

The first General Assembly of the International Biological Programme, attended by Representatives of the Scientific Academies of 30 nations and of 12 International Scientific Organisations and by a large number of the world's leading authorities on biological natural resources, has carefully considered and agreed upon a programme of Design and Feasibility studies which, if carried out, would lead to the definition of the research facilities and programmes considered most likely to contribute to the optimum exploitation, on a global basis, of the biological resources on which mankind is vitally dependent for its food and for many other products.

The Assembly resolves that this programme of Design and Feasibility studies can be carried out only if financial support is made available, by the national and/or international bodies which administer funds for scientific research, at an adequate level which may be estimated at a minimum of \$250 000 to \$500 000 per annum for two years, for its multi-national expenses.

From II General Assembly, Paris 1966

Co-operation with Specialist Agencies of UN

The following resolutions were proposed by SCIBP and were accepted unanimously by the General Assembly:

1. The SCIBP resolves to express its gratitude to UNESCO for its generosity in making available the excellent accommodation which has contributed greatly to the success of this Assembly, and also for the unstinting way in which the staff of the Organisation have made themselves available for consultation and advice.

2. The SCIBP resolves to express its deep appreciation of the co-operation of the specialised agencies of the United Nations in developing the International Biological Programme.

Appendix 1

From IV SCIBP, Paris 1967

Discrepancy between available food and quality of life

Biologists, as members of the Special Committee for the International Biological Programme, meeting in Paris for the planning phase of this programme, expressed grave concern with the increasing discrepancy between the availability of food and the dignity and quality of human life. Technological improvements to alleviate starvation, while hopeful, will not arrive in time to offset much human suffering and the long-lasting effects of malnutrition on the youth of the next generation.

These biologists counsel their colleagues throughout the world to offer their services in explaining the reality of population problems. They should also participate in arriving at intelligent decisions to improve human happiness, which depends on food, water and quality of the landscape.

From VI SCIBP, London 1969

1. SCIBP notes the proposals for the establishment by ICSU of a Special Committee on Problems of the Environment, welcomes the interest shown by ICSU in problems of the environment and urges that ICSU should take into account:

(*a*) the over-riding importance of biology in the study of the human environment;
(*b*) the large number of uncompleted international co-operative projects initiated in the IBP and co-ordinated by a network of international committees and some 60 national committees;
(*c*) the importance of continuation of the appropriate IBP-generated research under sponsorship of either intergovernmental or non-governmental bodies, or both;
(*d*) the importance of a positive and fully co-ordinated ICSU relation to intergovernmental programmes in biology.

2. SCIBP considers it most important that the report of the *ad hoc* Committee be critically examined by ICSU to identify activities appropriate to ICSU with due regard to the activities of other organisations.

3. Concerning a proposal that, in order to take full benefit from results, IBP should continue for a period after 1972, the viewpoint of national committees is divided and a fully co-ordinated recommendation cannot

be expected before the IBP General Assembly of October 1970. However, SCIBP requests ICSU to anticipate providing for not more than two years beyond 1972 for the completion and writing up of IBP projects now in progress.

From IV General Assembly, Rome 1970

Extension of the IBP

Noting that IBP has been successful in achieving one of its main objectives, namely the stimulation of programme co-operation in many branches of environmental biology;

Noting the desirability of ensuring at the end of IBP the continuation and development of this co-operation under other agencies in the best possible conditions which will entail synthesis of the results of IBP and often the transfer of continuing activities;

Noting that ICSU at its General Assembly, September 1970, has authorised that IBP should be extended after its scheduled ending in July 1972 for a further two years in order to complete its task effectively;

Agrees that this extension period, which follows Phase II (*Operational*) should be designated Phase III (*Synthesis and Transfer*) (synthesis of results so far obtained and transfer of responsibility for the co-ordination of selected activities to appropriate organisations);

Agrees further that during this period the essential structure of IBP will be maintained including the central and sectional organisation, in so far as this is necessary. Synthesis of results will need to continue for the full two years of Phase III and for some time thereafter. Transfer of continuing activities will not take place at a particular time but for some subjects will be possible early, for others later in Phase III according to the completion of arrangements with the international agencies concerned.

Transfer of Responsibility for Global Monitoring

Accepts the reports of the *ad hoc* committee; and

Agrees that SCIBP should transfer to SCOPE its responsibility for the consideration of matters relating to global monitoring;

Agrees further that SCIBP should seek the comments of its Sectional Committees on the *ad hoc* Committee's report (and appendices) and pass these on to SCOPE;

Appendix 1

Agrees in addition that SCIBP should suggest to SCOPE that the reports (and appendices) be sent to national adhering bodies of ICSU for comment and where appropriate to National IBP Committees.

Shortage of Environmental Scientists

Draws attention to the shortage of trained taxonomists, ecologists and other environmental scientists, that has been revealed during the operation of IBP, this shortage being largely due to inadequate training facilities at institutions of higher learning;

Emphasises the need to expand these facilities, to create the additional posts required, and to arrange for international exchange of students and university teachers in these fields to alleviate the shortage in the meantime; and

Expresses the desire that within the UN agencies steps be taken to establish international co-operation in:

(*a*) Environmental Education and Ecology Teaching (at schools and universities; curriculum reforms; ecological teaching guides);
(*b*) Training of Environmental Technicians and Ecologists (through courses, fellowships, visitorships, workshops);

and that more attention be given to the establishment, particularly in tropical and sub-tropical countries, of field centres which should be helped in equipment and staffing through the support of the UN agencies.

International Ecological Field Stations

Noting the requests of a number of Directors of ecological research stations for international recognition of their problems in developing biome studies in remote situations and for aid in improving their finances, their recruitment of personnel and their communications;

Requests the Convener of IBP/CT to pursue discussions with the Director-General of IUCN with the object of giving effect to these requests.

From V General Assembly, Seattle 1972

Relation between Man and his Environment in the Tropics

Considering the necessity of giving their proper place in the various forthcoming research programmes such as MAB, SCOPE, PAWE,

144

etc., to human biology and demography, epidemiology and allied sciences,

Considering the growing importance in developing tropical countries of the problems related to cross-culturisation, industrialisation, migration and transitional stages.

The V General Assembly of IBP Meeting in Seattle Recommends that, in the projects to be defined and promoted by the various organisations concerned with the transfer of IBP activities full attention be given to the effects of the environment on human communities as well as the effects of human communities on the environment.

Conservation of Wild Species and Genetic Diversity

Considering that special problems of conservation are presented by wild species of direct usefulness, such as many forest and grassland species and various mammals and fishes, and wild species of indirect or potential use, including the wild relatives of domesticated plants and animals, especially in the centres of genetic diversity;

Considering also that the objective must be the conservation not only of species as such, but also of an unusually wide range of genetic diversity, and that conservation within natural communities is the optimal measure for wild organisms.

The V General Assembly, having noted progress made in the course of IBP towards the conservation of the genetic resources of domesticated plants,

Recommends in general that important sites of species significant to human welfare be identified as a matter of urgency, bearing in mind the requirement of genetic diversity; that such sites be described and entered in National or International data banks; and that, where appropriate, steps be taken for their conservation in National Parks or equivalent areas, or in special reserves designed to conserve gene pools of particular species,

And recommends specifically to ICSU that studies of the biological problems and methodologies of conservation of the wild species, as delineated above, be stimulated, that survey be encouraged, and that conservation and registration measures be instituted in conjunction with UNESCO (Project No. 8 of MAB) and IUCN, and where appropriate with FAO (Forest, Grassland, Fishery and Weed Species) and IUFRO (Forest Species).

Appendix 1

The Need for Taxonomic Biology

Recognizing from the experience of IBP, that the effective development of ecological research requires a much greater number than at present of biologists trained in modern taxonomy and having an ecological outlook;

The V General Assembly of IBP

Recommends that ICSU should make this urgent need as widely known as possible, requesting that greater emphasis than at present be placed on taxonomy as part of the training of university biologists and as a research option;

That those budgeting for ecological research programmes should normally include sums for taxonomic services and taxonomic posts;

And that museums, herbaria, botanical and zoological gardens and national biological surveys should direct more of their research effort towards taxonomic aspects of the various environmental programmes now being planned.

Inter-Disciplinary Studies on the Environment

Noting with appreciation the growing interest of intergovernmental agencies in environmental activities,

Recognising the importance of the scientific community's contributions to these activities,

The V General Assembly of IBP

Recommends that ICSU should make a detailed survey of the ongoing and planned environmental activities of its Unions and Committees, encourage inter-disciplinary co-operation, both within ICSU and with other international organisations, and in its planning for future environmental activities ensure that the closest collaboration is established with other international bodies (governmental and non-governmental) that represent the total spectrum of environmental interests (biological, physical, and social sciences).

Future Research on Production Processes

Considering the fundamental significance of photosynthesis studies and of investigations on nitrogen fixation for ecological research in the future, and the good work performed in these subjects by the Section on Production Processes of the IBP,

146

The V General Assembly of the IBP

Invites the International Association of Plant Physiology (IAPP) to provide an organisational framework for perpetuation of these lines of work, and

Requests SCIBP to approach IAPP on this matter through the Secretariat of IUBS.

From IX SCIBP, London 1974

The Man and Biosphere Programme

Considering that much of the experience and knowledge acquired under IBP and many of the methodologies developed under its aegis constitute basic elements for the planning and implementation of the MAB programme The Special Committee for the IBP at its final meeting

Recommends that the Co-ordinating Council of MAB and the national IBP Committee in each country take all appropriate steps to ensure that this experience, this knowledge and these methodologies be fully taken into consideration by the National Committee for the MAB Programme of that country when formulating and implementing its participation in the international Programme.

International Co-operation in Biology

Considering that an important achievement of IBP has been the creation of an effective international network of co-operation and teamwork among biologists concerned with environment, human biology, and the processes of biological productivity; and that this achievement has been possible because of the flexibility and directness of a scientist to scientist approach across frontiers, The Special Committee for the IBP at its final meeting

Urgently recommends that ICSU encourage and support initiatives by IUBS and the other bioscience Unions to take responsibility for ensuring the continuance of a programme of international co-operation, including interdisciplinary exercises among research biologists, and explore ways in which this can be funded, and respond sympathetically to proposals from former IBP working groups.

Membership of SCIBP and its main committees

Officers and members of SCIBP as at 30 June 1974
(The year of appointment follows each name)

President Emeritus
J. G. Baer (Switzerland) 1964 (died 1975)

Officers

President
F. Bourlière (France) 1969

Vice-Presidents
W. F. Blair (USA) 1969
Sir Otto Frankel (Australia) 1966
I. Málek (Czechoslovakia) 1969
H. Tamiya (Japan) 1967

Scientific Director
E. B. Worthington (UK) 1964

Members

Conveners of Sectional Committees

PF	J. B. Cragg (Canada) 1969
PP	I. Málek (Czechoslovakia) 1969
CT	E. M. Nicholson (UK) 1964
PF	L. Tonolli (Italy) 1970
PM	M. J. Dunbar (Canada) 1970
HA	J. S. Weiner (UK) 1964
UM	G. K. Davis (USA) 1964

Representing ICSU, its member unions and scientific committees

ICSU	F. B. Straub (Hungary) 1971
IUBS	A. D. Hasler (USA) 1970
IUB	M. Florkin (Belgium) 1964

Appendix 2

IUPS	J. S. Weiner (UK) 1964
IGU	K. M. Clayton (UK) 1970
IUPAB	A. A. Gopal-Ayengar (India) 1964
IUNS	C. G. King (USA) 1967
SCAR	G. A. Knox (New Zealand) 1969
SCOR	R. C. Dugdale (IABO, USA) 1972

Representing other international organisations

IUCN	J. G. Baer (Switzerland) 1969
IUAES & IAHB	J. Hiernaux (Belgium) 1966

Elected members

W. F. Blair (USA) 1968
F. Bourlière (France) 1969
A. E. Boyo (Nigeria) 1966
B. Bychowsky (USSR) 1964 (died 1974)
Sir Otto Frankel (Australia) 1969
E. R. Hajek (Chile) 1972
R. W. J. Keay (UK) 1964
G. Montalenti (Italy) 1964
K. Petrusewicz (Poland) 1964
F. Salzano (Brazil) 1969
B. R. Seshachar (India) 1968
H. Tamiya (Japan) 1964
H. Thamdrup (Denmark) 1968
C. A. du Toit (South Africa) 1966

Past Officers and members of SCIBP

Officers

President

J. G. Baer (Switzerland) 1964–9

Vice-Presidents

S. A. Cain (USA) 1964–7
G. Montalenti (Italy) 1964–9
K. Petrusewicz (Poland) 1964–9
C. H. Waddington (UK) 1964–6

Members

Conveners of Sectional Committees

PT	F. Bourlière (France) 1964–9
PF	V. Tonolli (Italy) 1964–6
	A. Hasler (USA) 1966–8
	G. G. Winberg (USSR) 1963–70
PM	R. S. Glover (UK) 1964–6
	B. H. Ketchum (USA) 1966–8

Representing

ICSU	D. Blaskovic (Czechoslovakia) 1964–70
IUBS	C. H. Waddington (UK) 1964–9
IGU	C. Troll (FDR) 1964–70
IUNS	D. P. Cuthbertson (UK) 1964–7
IUPS	S. Robinson (USA) 1966
	P. O. Astrand (Sweden)
SCAR	M. W. Holdgate (UK) 1966–8
SCOR	O. H. Oren (Israel) 1966–72

Elected members

A. A. Gopal-Ayengar (India) 1964–8
S. A. Cain (USA) 1964–9
C. G. Heden (Sweden) 1964–8
M. Hyder (Kenya) 1964
M. Vannucci (Brazil) 1966–9
F. di Castri (Chile) 1969–72

Finance committee
(established by SCIBP, July 1964) as at 30 June 1974

Chairman
R. W. J. Keay (UK)

Members
C. A. du Toit (South Africa) 1969
N. B. Cacciapuoti (ICSU)

Adviser
Scientific Director

Appendix 2

Past members

T. C. Byerly (USA) 1964–7
R. Revelle (USA) 1966–8
W. F. Blair (USA) 1968–9
G. R. Laclavère (ICSU) 1964

Editorial committee
for International Synthesis (established by SCIBP in March 1972)

The President and Scientific Director
G. E. Blackman (UK)
P. T. Baker (USA)
G. G. Winberg (USSR)

Staff of Central Office

Scientific Director
   E. B. Worthington 1964
Executive Secretary
   H. A. W. Southon 1965–73
   Gina Douglas (from 1973)
Executive Officer
   Sue Darell-Brown (from 1971)
Executive Assistant
   Sheila Alderson 1967–70

Sectional Committees

The following have served for part or all of the period as members of sectional committees or as international leaders of themes:

PT: Productivity of Terrestrial Communities

Balcells, H. (Spain)
Blaxter, K. L. (UK)
Bliss, L. C. (Canada)
Bourlière, F. (France), (Convener)
Brian, M. V. (UK)
Brown, J. (USA)
Cantlon, J. E. (USA)
Coupland, R. T. (Canada)
Cragg, J. B. (Canada), (Convener)

Delamare Deboutteville, C. (France)
Douglas, G. (Co-ordinator, part-time)
Duvigneaud, P. (Belgium)
Ellenberg, H. (BRD)
Elliott, H. F. I. (Co-ordinator, part-time)
Evenari, M. (Israel)

152

PT: Productivity of Terrestrial Communities – continued

François, A. (France)
Furusaka, C. (Japan)
Ghilarov, M. S. (USSR)
Giacomini, V. (Italy)
Goodall, D. W. (USA)
Hadley, M. J. (Co-ordinator)
Heal, O. W. (UK)
Kira, T. (Japan)
Kühnelt, W. (Austria)
Lemée, G. (France)
Lieth, H. (BDR)
Monsi, M. (Japan)

Odum, E. P. (USA)
Ovington, J. D. (Australia)
Parkinson, D. (Canada)
Perry, R. A. (Australia)
Petrusewicz, K. (Poland)
Phillipson, J. (UK)
Pinowski, J. (Poland)
Reichle, D. E. (USA)
Schwartz, S. S. (USSR)
Talbot, L. M. (USA)
Treccani, V. (Italy)
Varley, G. C. (UK)

PP: Production Processes

Abd-el-Malek, Y. (Egypt)
Alexander, M. (USA)
Allen, O. N. (USA)
Anderson, M. C. (Australia)
Bergersen, F. J. (Australia)
Blackman, G. E. (UK)
Bond, G. (UK)
Bourdu, R. (France)
Burström, H. (Sweden)
Cooper, J. P. (UK)
Delwiche, C. C. (USA)
Eckardt, F. E. (France)
Fahraeus, G. (Sweden)
Fogg, G. E. (UK)
Gaastra, P. (Netherlands)
Hamatová, E. (Czechoslovakia) (Co-ordinator: PP-N)
Ishizawa, S. (Japan)
Jansson, S. (Sweden)
Jenkinson, D. S. (UK)
Jensen, H. L. (Denmark)

Květ, J. (Czechoslovakia) (Co-ordinator: PP-P)
Lemon, E. R. (USA)
Lie, T. A. (Netherlands)
Málek, I. (Czechoslovakia) (Convener)
Mishustin, E. N. (USSR)
Monsi, M. (Japan)
Monteith, J. L. (UK)
Moyse, A. (France)
Mulder, E. G. (Netherlands)
Müller, D. (Denmark)
Nathorst-Westfelt, L. (Sweden)
Nichiporovich, A. A. (USSR)
Nutman, P. S. (UK)
Pate, J. S. (UK)
Quispel, A. (Netherlands)
Reifer, I. (Poland)
Saeki, T. (Japan)
Schwartz, W. (BRD)
Senez, J. C. (France)

Appendix 2

PP: Production Processes – continued

Setlik, I. (Czechoslovakia)

Slavík, B. (Czechoslovakia)

Spaldon, E. (Czechoslovakia)

Specht, R. L. (Australia)

Stewart, W. D. P. (UK)

Talling, J. F. (UK)

Vincent, J. M. (Australia)

Wassink, E. L. (Netherlands)

Wilson, P. W. (USA)

Zalensky, O. V. (USSR)

CT: Conservation of Terrestrial Communities

Acosta-Solis, M. (Ecuador)

Bannikov, A. G. (USSR)

Budowski, G. (Costa Rica)

Clapham, A. R. (UK)

Dasmann, R. (Switzerland)

Dementiev, G. P. (USSR)

Dorst, J. (France)

Douglas, G. (UK) (Co-ordinator)

Fosberg, F. R. (USA)

Goetel, W. (Poland)

Graham, E. H. (USA)

Holdgate, M. W. (UK)

Kassas, M. (Egypt)

Medwecka-Kornas, A. (Poland)

Monod, Th. (France)

Mueller-Dombois, D. (USA)

Nicholson, E. M. (UK) (Convener)

Peterken, G. F. (UK) (Co-ordin-ator)

Poore, M. E. D. (UK)

Radford, G. L. (UK)

Talbot, L. M. (USA)

Valverde, J. A. (Spain)

Vtorov, P. (USSR)

Webb, L. (Australia)

PF: Productivity of Freshwater Communities

Allsopp, H. (British Honduras)

Bonetto, A. A. (Argentina)

Evens, F. (Belgium)

Ganapati, S. V. (India)

Gerking, S. (USA)

Goldman, C. (USA)

Golterman, H. (Netherlands)

Hasler, A. (USA) (Convener)

Hrbacek, J. (Czechoslovakia)

Hynes, N. (Canada)

Jónasson, S. (Denmark)

Kajak, Z. (Poland)

Le Cren, E. D. (UK)

Lund, J. (UK)

Mann, K. H. (Canada)

Marlier, G. (Belgium)

Mori, S. (Japan)

Patalas, K. (Poland)

Rodhe, W. (Sweden)

Rzóska, J. (UK) (Co-ordinator)

Sioli, H. (BDR)

Tonolli, L. (Italy) (Convener)

Tonolli, V. (Italy) (Convener)

Vibert, R. (France)

PF: Productivity of Freshwater Communities – continued

Watt, K. (USA)
Williams, W. D. (Australia)

Winberg, G. G. (USSR) (Convener)

PM: Productivity of Marine Communities

Angot, M. (Madagascar)
Bayne, B. L. (UK)
Beklemisher, K. V. (USSR)
Brock, V. (USA)
Cassie, R. M. (New Zealand)
Clark, R. B. (UK)
Crisp, D. J. (UK)
Dohrn, P. (Italy)
Doty, M. S. (USA)
Dugdale, R. C. (USA)
Dunbar, M. J. (Canada) (Convener)
Filatova, Z. A. (USSR)
Glover, R. (UK) (Convener)
Harrison Matthews, L. (UK)

Hogetsu, K. (Japan)
Humphrey, G. (Australia)
Ketchum, B. H. (USA) (Convener)
Krey (BDR)
Mann, K. H. (Canada)
Margalef, R. (Spain)
Olaniyan, C. I. O. (Nigeria)
Oren, O. H. (Israel)
Pearce, J. B. (USA)
Ray, C. (USA)
Soegiarto, A. (Indonesia)
Vannucci, M. (Brazil)
Walford, L. A. (USA)
Yonge, C. M. (UK) (Co-ordinator)

HA: Human Adaptability

Andjus, J. (Yugoslavia)
Baker, P. T. (USA)
Barbashova, Z. I. (USSR)
Bielicki, T. (Poland)
Boyo, A. E. (Nigeria)
Carlson, L. D. (USA)
Cavalli-Sforza, L. L. (Italy)
Collins, K. (UK) (Co-ordinator)
Cotes, J. E. (UK)
Fox, R. H. (UK)
Garrad, G. (UK) (Co-ordinator)
Harrison, G. A. (UK)
Hart, S. (Canada)
Hiernaux, J. (Belgium)
Hulse, F. S. (USA)

Ketusinh, O. (Thailand)
King, C. G. (USA)
Kostial, K. (Yugoslavia)
Lange-Andersen, K. (Norway)
Lourie, J. (UK) (Co-ordinator)
Macek, M. (Czechoslovakia)
Malhotra, M. S. (India)
Milan, F. A. (USA)
Mourant, A. E. (UK)
Musgrave, J. (UK) (Co-ordinator)
Neel, J. V. (USA)
Poupa, O. (Czechoslovakia)
Robinson, S. (USA)
Salzano, F. M. (Brazil)
Sanghvi, L. D. (India)

155

Appendix 2

HA: Human Adaptability – continued

Sutter, J. (France)

Tanner, J. M. (UK)

Tobias, P. V. (South Africa)

Walsh, R. J. (Australia)

Wanke, A. (Poland)

Weiner, J. S. (UK) (Convener)

Wyndham, C. H. (South Africa)

Yoshimura, H. (Japan)

UM: Use and Management of Biological Resources

Bateman, M. A. (Australia)

Benninghoff, W. S. (USA)

Blaxter, K. L. (UK)

Cuthbertson, D. (UK)

Davis, G. K. (USA) (Convener)

DeBach, P. (USA)

Delucchi, V. (Switzerland)

Donald, C. M. (Australia)

Finlay, K. W. (Mexico)

Frankel, O. H. (Australia)

Grison, P. (France)

Hedén, C.-G. (Sweden)

Huffaker, C. B. (USA)

Hussey, N. W. (UK)

Hyder, M. (Kenya)

Mackauer, M. (Canada)

Matsuo, T. (Japan)

Murty, B. R. (India)

Peterson, R. A. (FAO)

Pirie, N. W. (UK)

Stebbins, G. C. (USA)

Swaminathan, M. S. (India)

Tannenbaum, S. R. (USA)

Yasamatsu, K. (Japan)

156

APPENDIX 3

National organisation of the IBP

3a. National participation

ARGENTINA

Adhering Organisation: Consejo Nacional de Investigaciones Cientificas y Técnicas, Rivadavia 1917–R25, Buenos Aires.
Chairman of IBP Committee: Prof. O. Boelcke. *Sub-committees* for PT/CT, PF, PM, HA, UM.

AUSTRALIA

Adhering Organisation: Australian Academy of Science, Gordon Street, Canberra City, 2601, A.C.T.
Chairman of IBP Committee: Sir Otto Frankel. *Secretary:* J. Deeble. *Sub-committees* for all sections.

AUSTRIA

Adhering Organisation: Die Osterreichische Akademie der Wissenschaften, Dr Ignaz Seipel-Platz 2, Wien 1.
Chairman of IBP Committee: Dr W. Kühnelt. *Sub-committees* for PT, PP, PF, HA.

BELGIUM

Adhering Organisation: Académie Royale de Belgique, Palais des Académies, Bruxelles.
Chairman of IBP Committee: Prof. P. Duvigneaud.
Secretary: Dr G. Marlier. *Sub-committees* for PT, PP, CT, PF, HA, UM.

BRAZIL

Adhering Organisation: Conselho Naçional de Pesquisas, Avenida Marechal Camara 350, Rio de Janeiro-GB.
National Committee and Sub-committees for PT/PP, CT, PF/PM, HA, UM.

BULGARIA

Adhering Organisation: Bulgarian Academy of Sciences, 7 November-Street, No. 1, Sofia.
Secretary of National Committee: Prof. K. I. Markov. *Sub-committees* for PT/CT, PP, PF/PM, HA, UM.

Appendix 3

Adhering Organisation: National Research Council of Canada, Montreal Road, Ottawa 7, Ontario.
Chairman of IBP Committee: Dr T. W. M. Cameron. *Director-General:* Dr W. H. Cook. *Sub-committees* for all sections.

CHILE
Adhering Organisation: National Research Council of Chile, CONICYT, Casilla 297-V, Santiago.
Chairman of IBP Committee: Dr Ernst R. Hajek. *Sub-committees* for PT, CT, PF, PM, HA.

COLUMBIA
Adhering Organisation: Academia Colombiana de Ciencias Exactas, Fisicas y Naturales, Apt. 2584, Bogota.
Chairman of IBP Committee: Dr Luis Eduardo Mora.

CZECHOSLOVAKIA
Adhering Organisation: Ceskoslovenská Akademie Ved., Narodni Tr. 3, Praha 1.
Chairman IBP Committee: Prof. Dr E. Hadač. *Secretary:* J. Hrbacek. *Sub-committees* for PT, PP, CT, PF, HA, UM.

DENMARK
Adhering Organisation: Akademiet for de Tekniske Videnskaber, ATV, Rigensgade 11, DK 1316, København K.
Chairman of IBP Committee: Prof. H. M. Thamdrup. *Secretary:* H. Peterson. *Sub-committees* for PT, PF, UM.

EAST AFRICA
Adhering Organisation: East African Academy, P.O. Box 7288, Nairobi, Kenya.
(See also entries under Kenya, Tanzania and Uganda.)

ECUADOR
Adhering Organisation: Instituto Ecuatoriano de Ciencias Naturales. Main Office: Calle Manabí 568 (3er piso), Apartado 408, Quito.
Chairman of IBP Committee: Dr Misael Acosta-Solis. *Secretary:* A. Bustamante. *Sub-committees* for all sections.

EGYPT
Adhering Organisation: Ministry of Scientific Research, UAR Department of Scientific Societies and International Unions, 101 Kasr El-Einy Street, Cairo.
Chairman of IBP Committee: Prof. Hussein Said. *Secretary:* M. Hafez.

FINLAND
Adhering Organisation: Suomalainen Tiedeakatemia, Snellmaninkatu 9–11, Helsinki 17.
Chairman of IBP Committee: Prof. Hans Luther. *Secretary:* C. A. Haeggstrom. *Sub-committees* for all sections.

FRANCE
Adhering Organisation: Académie des Sciences, Institut de France, 23 Quai Conti, Paris 6ᵉ.
Chairman of IBP Committee: Prof. Th. Monod. *Secretary:* P. Grison. *Sub-committees* for all sections.

DEMOCRATIC REPUBLIC OF GERMANY (DDR)
Adhering Organisation: Deutsche Akademie der Wissenschaften zu Berlin, 1080 Berlin, Otto-Nuschke-Strasse 22–23.
Chairman of IBP Committee: Prof. Dr H. Stubbe. *Sub-committees* for all sections.

FEDERAL REPUBLIC OF GERMANY (BRD)
Adhering Organisation: Deutsche Forschungsgemeinschaft,Kennedyallee 40, D-5300 Bonn–Bad Godesberg.
Chairman of IBP Committee: Prof. Dr H. Ellenberg. *Secretary:* Dr H. Heller. *Sub-committees* for all sections.

GHANA
Adhering Organisation: Ghana Academy of Sciences, P.O. Box M.32, Accra.
Sub-committees for PT/CT/UM, PF/PM, HA.

GREECE
Adhering Organisation: Akadimia Athinon, Odos Panepistimiou, Athens.

Appendix 3

HUNGARY
Adhering Organisation: Hungarian Academy of Sciences, Department of Biology, Budapest V, Nádor u. 7.
Chairman of IBP Committee: Prof. J. Balogh. *Secretary:* Dr G. Vida.
Sub-committees for PT, PP, CT, PF, HA.

INDIA
Adhering Organisation: Indian National Science Academy, Bahadur Shah Zafar Marg, New Delhi.
Chairman of IBP Committee: Prof. B. R. Seshachar. *Sub-committees* for PT, CT, PF, PM, HA, UM.

INDONESIA
Adhering Organisation: Lembaga Ilmu Pengetahuan Indonesia, Djalan Teuku Tjhik Ditiro 43, Djakarta.
Chairman of IBP Committee: Dr Otto Soemarwoto. *Sub-committees* for all sections.

REPUBLIC OF IRELAND
Adhering Organisation: Royal Irish Academy, 19 Dawson Street, Dublin 2.
Chairman of IBP Committee: Mr P. M. McDonnell. *Secretary:* Fr J. J. Moore. *Sub-committees* for PT, PP, CT, UM.

ISRAEL
Adhering Organisation: Israel Academy of Sciences and Humanities, P.O. Box 4040, Jerusalem.
Chairman of IBP Committee: Prof. M. Evenari. *Secretary:* Dr H. Ginsburg. *Sub-committees* for PT/PP, CT, PM/PF, HA, UM.

ITALY
Adhering Organisation: Consiglio Nazionale delle Ricerche, Piazzale delle Scienze 7, 00185, Roma.
Chairman of IBP Committee: Prof. Claudio Barigozzi. *Sub-committees* for all sections.

JAPAN
Adhering Organisation: Science Council of Japan, 22–34 Roppongi 7 Chome, Minato-ku, Tokyo 106.
Chairman of IBP Committee: Prof. H. Tamiya. *Secretary:* M. Monsi. *Sub-committees* for all sections.

KENYA
Adhering Organisation: East African Academy, P.O. Box 7288, Nairobi.
Chairman of IBP Committee: Dr R. S. Odingo.

REPUBLIC OF KOREA
Adhering Organisation: National Academy of Sciences, 1, Senjongro, Jongro-gu, Seoul.
Chairman of IBP Committee: Prof. Yung-sun Kang. *Secretary:* Dr Chung Choo Lee. *Sub-committees* for PT/PP/UM, CT, PF, HA.

MALAWI
Chairman of IBP Committee: Dr N. P. Mwanza, Chancellor College, University of Malawi, P.O. Box 5200, Limbe. *Secretary:* M. Kalk.

MALAYSIA
Chairman of IBP Committee: Enche K. D. Menon, Pusat Penyelidekan Hutan, Kepong, Selangor. *Sub-committees* for PT, CT, PF, HA.

MEXICO
Adhering Organisation: Academia de la Investigación Científica. CONACYT, Insurgentes sur 1814, Mexico 20, D.F.
Co-Chairmen of IBP Committee: Dr Agustin Ayala-Castanares, Dr Alfredo Barrera. *Secretary:* Dr Raúl Macgregor. *Sub-committees* for PT, PP, CT, PF/PM, HA, UM.

NETHERLANDS
Adhering Organisation: Royal Netherlands Academy of Sciences and Letters, Kloveniersburgwal 29, Amsterdam – C.
Chairman of IBP Committee: Prof. G. J. Vervelde. *Sub-committees* for all sections.

NEW ZEALAND
Adhering Organisation: Royal Society of New Zealand, Box 12249, Wellington.
Chairman of IBP Committee: Prof. G. A. Knox. *Sub-committees* for PT/PP, CT, PF/PM, HA.

NIGERIA
Adhering Organisation: Science Association of Nigeria, School of Biological Sciences, University of Lagos, Lagos.
Chairman of IBP Committee: Prof. C. I. O. Olaniyan. *Secretary:* Ola Ojikutu. *Sub-committees* for all sections.

Appendix 3

NORWAY

Adhering Organisation: Norwegian Research Council for Science and the Humanities, Wergelandsveien 15, Oslo 1.
Chairman of IBP Committee: Prof. Rolf Vik. *Secretary:* Dr F. E. Wielgolaski. *Sub-committees* for all sections.

PANAMA

Chairman of IBP Committee: Dr Octavio E. Sousa, Escuela de Biologia, Universidad de Panamá, P.O. Box 6403, Panamá 5. *Secretary:* Dra Reina Torres de Arauz.

PERU

Adhering Organisation: Museo de Historia Natural de la Universidad Naçional, Mayor de San Marcos, Apartedo 1109, Lima.
Chairman of IBP Committee: Prof. Hernando de Macedo.

PHILIPPINES

Adhering Organisation: National Research Council of the Philippines, University of the Philippines, Diliman, Rizal.
Chairman of IBP Committee: Dr T. Velasquez. *Secretary:* Prof. Rogelio O. Juliano. *Sub-committees* for all sections.

POLAND

Adhering Organisation: Polish Academy of Sciences, Warsaw, Palace of Culture, P.O. Box 2603.
Chairman of IBP Committee: Prof. K. Petrusewicz. *Secretary:* Dr A. Andrzejewska. *Sub-committees* for PT/CT, PP, PF, PM, HA.

RHODESIA

Adhering Organisation: Scientific Council of Rhodesia.
Chairman of IBP Committee: Prof. E. Bursell. *Sub-committees* for PT, PP, CT, PF, HA, UM.

ROMANIA

Adhering Organisation: Romanian Academy of Sciences, Splaiul Independentei 296, Raion 16 Februarie, Bucharest.
Chairman of IBP Committee: Prof. N. Salageanu. *Secretary:* Prof. R. Codreanu. *Sub-committees* for PF, PM, HA, UM.

162

SOUTH AFRICA

Adhering Organisation: South African Council for Scientific and Industrial Research, P.O. Box 395, Pretoria.
Chairman of IBP Committee: Prof. C. A. Du Toit. *Secretary:* Mr R. G. Noble. *Sub-committees* for all sections.

SPAIN

Adhering Organisation: Conseho Superior de Investigaciones Cientificas.
Chairman of IBP Committee: Prof. A. Carrato. *Secretary:* Dr E. Balcells. Centro Pirenaico de Biologia Experimental B.P. 64 Jaca (Prov. of Huesca).
Sub-committees for PT/PP, CT, PF/PM, HA, UM.

SRI LANKA

Adhering Organisation: Sri Lanka Association for the Advancement of Science, 281/50 Bauddhaloka Mawatha, Colombo 7.
Chairman of IBP Committee: Prof. B. A. Abeywickrama. *Secretary:* Dr P. Canagaratnam. *Sub-committees* for PT/CT, PF, PM.

SWEDEN

Adhering Organisation: Kungl. Vetenskaps Akademien.
Chairman of IBP Committee: Prof. Carol Olof Tamm. *Secretary:* Dr T. Rosswall, Wenner-Gren Center, Sveavägen, 166 8tr. S-113 46 Stockholm. *Sub-committees* for all sections.

TAIWAN

Adhering Organisation: Academia Sinica, 130 Yen Chiu Yuan Road Section 1, Nankang, Taipei, Taiwan.
Chairman of IBP Committee: Dr Jong-Ching Su. *Sub-committees* for all sections.

TANZANIA

Adhering Organisation: The East African Academy (Tanzania Branch), Box 35033, University Hill, Dar es Salaam.
Chairman of IBP Committee: Dr A. S. Msangi. *Secretary:* Miss A. McCusker. *Sub-committees* for CT, PM, HA.

THAILAND

Adhering Organisation: National Research Council of Thailand, Bangkhen, Bangkok 9.
Chairman of IBP Committee: Mr Insee Chandrastitya. *Secretary:* Dr P. Cheosakul. *Sub-committees* for all sections.

UGANDA

Adhering Organisation: East African Academy (Uganda Branch) P.O. Box 7062, Kampala.

UK

Adhering Organisation: The Royal Society, 6, Carlton House Terrace, London, SW1Y 5AG.
Chairman of IBP Committee: Prof. A. R. Clapham. *Secretary:* Mrs. R. Z. Bulsara.

USA

Adhering Organisation: National Academy of Sciences, 2101 Constitution Avenue, Washington DC 20418.
Chairman of IBP Committee: Dr J. F. Reed. *Staff Officer:* R. H. Oliver.

USSR

Adhering Organisation: USSR Academy of Sciences, Leninsky Prospect 14, Moscow B.71.
Chairman of IBP Committee: Prof. M. S. Gilarov. *Vice-Chairman* Dr O. N. Bauer. *Sub-committees* for PT, PP, CT, PM, HA, UM.

URUGUAY

Chairman of IBP Committee: Dr C. Estable, c/o UNESCO Regional Centre, Casilla de correo 859, Montevideo. *Secretary:* Dr C. Batthyany.

VENEZUELA

Adhering Organisation: Academia de Ciencias Fisicas, Matemáticas y Naturales, Asociación Venezolana para el Avance de la Ciencia, Sociedad Venezolana de Ciencias Naturales, Caracas.
Chairman of IBP Committee: Dr M. Layrisse. *Sub-committees* for PT/PP, PF, PM, HA, UM.

REPUBLIC OF VIETNAM

Chairman of IBP Committee: Mr Pham Hoàng-Ho, Biological Society of Vietnam, Faculty of Sciences, B.P.A./2, Saigon. *Sub-committee* for PF.

YUGOSLAVIA

Adhering Organisation: Union of Yugoslav Scientific Biological Societies. c/o University of Zagreb, P.O. Box 327, 41000, Zagreb.
Chairman of IBP Committee: Prof. Dj. Jelenic. *Secretary:* Dr M. Todorovic. *Sub-committees* for all sections.

ZAIRE

Adhering Organisation: Office National de la Recherche et du Développement (ONRD), Kinshasa.
Chairman of IBP Committee: J. Ileo. *Secretary:* Prof. A. F. de Bont.
Sub-committees for PT, PP, PF, HA.

Other participating countries

The following forty countries, in which national committees have not been established, have been in touch with IBP through correspondents or have contributed projects to the programme through the participation of individual scientists or groups of scientists.

Algeria	Lebanon
Barbados	Libya
Bolivia	Malagasy
Botswana	Morocco
Burma	Mozambique
Burundi	Nepal
Central African Republic	Pakistan
Costa Rica	Papua
Côte d'Ivoire	Portugal
Cuba	Senegal
Ethiopia	Sierra Leone
Fiji	Singapore
Guatemala	Sudan
Haute Volta	Surinam
Hong Kong	Switzerland
Iran	Tongo
Iraq	Trinidad
Jamaica	Tunisia
Jordan	Turkey
Kuwait	Zambia

3b. Payment of national dues (shown in US dollars) due to SCIBP by participating countries

Country	1964	1965	1966	1967	1968	1969	1970*	1971*	1972*	1973*	1974*
Argentina	1 000	1 000	—	500	500	500	500	500	500	500	500
Australia	—	100	3 000	3 000	3 000	5 000	5 000	5 000	5 000	5 000	5 000
Austria	—	—	100	100	100	1 000	1 000	500	500	500	500
Belgium	—	—	—	1 000	1 000	1 000	1 000	1 000	1 000	1 000	(1 000)
Brazil	—	—	—	—	500	500	500	500	500	500	500
Bulgaria	—	100	100	100	100	500	500	500	500	500	500
Canada	—	2 500	2 500	4 000	4 000	7 500	7 500	7 500	7 500	7 500	3 750
Chile	—	—	—	—	—	—	—	(500)	(500)	—	—
Colombia	—	—	—	—	—	500	500	—	(500)	(500)	(500)
Czechoslovakia	—	200	200	200	200	500	500	500	500	500	500
Denmark	—	—	—	—	500	500	500	500	500	500	500
Ecuador	200	—	—	—	—	—	—	—	—	—	—
Finland	—	200	500	1 000	1 100	1 100	1 100	1 100	1 100	1 100	550
France	1 000	5 000	5 000	5 000	5 000	10 000	10 000	10 000	10 000	—	—
Germany, Fed. Rep.	—	1 000	1 000	1 000	1 000	1 000	1 000	1 000	1 000	—	—
Germany, Dem. Rep.	—	500	500	500	500	500	500	500	500	—	—
Ghana	—	—	—	—	—	—	—	—	—	—	—
Greece	—	100	100	100	100	100	100	—	500	(500)	—
Hungary	—	—	—	200	200	300	300	200	500	500	500
India	—	—	100	100	100	500	300	500	500	500	(500)
Indonesia	—	100	100	100	100	100	100	500	500	500	(500)
Ireland, Rep. of	—	—	—	100	500	500	500	500	500	(500)	(500)
Israel	100	100	500	500	500	500	500	500	500	500	(500)
Italy	—	5 000	5 000	5 000	5 000	5 000	5 000	5 000	5 000	5 000	5 000
Japan	—	—	2 500	2 500	2 500	2 500	5 000	5 000	5 000	5 000	5 000
Kenya	—	—	—	—	—	—	—	—	(200)	—	—
Korea	—	—	—	—	—	—	200	200	125	—	—
Malawi	—	—	—	—	—	—	—	—	500	500	—
Malaysia	—	—	—	—	—	500	500	500	500	500	500

Mexico	—	—	—	—	—	—	500	(500)	(500)	—	—
Netherlands	—	—	1 000	1 000	1 000	1 100	1 100	1 500	1 100	1 100	2 250
New Zealand	—	100	100	100	100	500	500	500	500	500	500
Nigeria	—	100	500	500	100	200	100	(100)	(100)	—	—
Norway	—	100	—	—	1 000	2 000	2 000	2 000	2 000	2 500	2 500
Panama	—	—	—	—	—	500	(500)	(500)	(500)	—	—
Peru	—	—	—	—	—	—	—	—	—	—	—
Philippines	—	—	—	—	100	500	500	500	500	500	500
Poland [See Note 2]	—	—	—	—	—	—	—	—	—	—	—
Rhodesia	—	100	100	100	100	500	500	500	500	500	500
Romania	—	—	—	—	—	200	500	500	500	500	500
South Africa	—	500	500	500	1 000	1 000	1 000	1 000	1 000	1 000	—
Spain	100	100	100	500	500	500	500	500	500	500	500
Sri Lanka	—	—	—	—	—	500	500	500	(500)	(500)	—
Sweden	2 500	5 000	5 000	5 000	5 000	5 500	5 500	5 500	5 500	5 500	2 750
Taiwan	—	100	100	100	100	100	500	500	500	500	500
Tanzania	—	100	—	—	—	167	(167)	(167)	(167)	—	—
Thailand	—	—	—	100	100	500	500	500	500	(500)	—
UAR	—	—	—	—	100	500	500	500	500	500	500
Uganda	—	—	—	—	—	—	—	—	—	—	500
United Kingdom	5 000	10 000	10 000	10 000	10 000	20 000	20 000	20 000	20 000	20 000	20 000
United States	5 000	10 000	10 000	10 000	10 000	17 500	25 000	25 000	25 000	25 000	12 500
USSR	—	—	—	2 500	2 500	5 000	7 000	7 000	7 000	7 000	7 000
Uruguay	—	—	—	—	—	—	—	—	—	—	—
Venezuela	—	—	—	—	250	500	(500)	(500)	(500)	—	—
Vietnam, Rep. of	—	—	—	—	—	500	500	(500)	(500)	—	—
Yugoslavia	100	100	100	100	100	100	500	500	500	500	500
Zaire	—	—	—	—	—	500	(500)	500	(500)	(500)	(500)
Received	14 900	41 500	48 800	55 600	58 050	97 467	110 000	109 800	109 325	96 200	74 800
Outstanding							1 667	2 767	3 967	3 500	4 000
Total	14 900	41 500	48 800	55 600	58 050	97 467	111 667	112 567	113 292	99 700	78 800

* Items in parentheses are payments outstanding at 4 March 1975.

[continued

3b. – continued

Notes

(1) In the table, national dues are shown under the years to which they relate, but may have been paid in a later year. The totals shown do not, therefore, correspond with those shown in the audited accounts, in which contributions are shown as income as and when received.

(2) The contributions shown include only those sums received into the SCIBP bank accounts in convertible currency. Many other contributions are made by participating countries in the form of contributions towards the cost of meetings and travel, the expenses of scientific co-ordinators, publications and other sectional activities. Particular mention may be made of the additional special contribution of $50 000 per annum in the years 1965–7 by the US National Academy of Sciences, the provision of office accommodation for the IBP Central Office byThe Royal Society of London since 1965, the provision of a scientific co-ordinator and office accommodation for Section CT by the National Environment Research Council of the United Kingdom since 1964, provision of scientific co-ordination for Section PP by the Czechoslovak Academy of Sciences since 1965, the provision of scientific co-ordination for Section PP by the Czechoslovak Academy of Sciences and support for meetings at the rate of approximately $2 500 per annum (1965–74) in non-convertible currency by the Polish Academy of Sciences. In addition, substantial contributions towards the expenses of the III General Assembly, Varna (1968), were arranged through the Bulgarian National Committee for IBP, and the printing and distribution of *IBP News*, no. 12 was undertaken by the Hungarian Academy of Sciences without cost to SCIBP.

The meetings of IBP

4a. Symposia and technical meetings, 1964–74

1964

Biology of human adaptability (HA/Wenner-Gren Foundation). Burg Wartenstein, Austria.

1965

April–May

Symposium on methods of studying primary productivity in fresh waters (PF). Pallanza, Italy.

July

Meeting on natural selection and transmissible disease (UK/HA). London, UK.

Functioning of terrestrial ecosystems at primary production level (UNESCO in association with PT and PP). Copenhagen, Denmark.

September

Large wild and tame herbivores (PT/UM in association with IUCN). Aberdeen and Cambridge, UK.

International symposium on environmental physiology (HA/IUPS/FASEB), Kyoto, Japan.

Symposium on human adaptability to environmental conditions and physical fitness (HA/DIPAS). New Delhi, India.

September–October

Symposium on man-made lakes (with Institute of Biology) (PF). London, UK.

November–December

Conservation in tropical SE Asia (CT). Bangkok, Thailand.

1966

March–April

Requirements and methods in ecological survey (CT). Monks Wood, UK.

July
Contributions to a world conservation programme (CT). Brunnen, Switzerland.

August
Working group on novel protein sources (UM). Warsaw, Poland.
Working group on biological control (UM). Tokyo, Japan.

August–September
Principles and methods of secondary productivity studies of terrestrial ecosystems (PT). Warsaw, Poland.

September
The biological basis of freshwater fish production (PF). Reading, UK.

October
Chemical methods in freshwater productivity studies (PF). Amsterdam, Netherlands.

Working group on plant gene pools (UM). Rome, Italy.

Working group on chemical methods (PM). Amsterdam, Netherlands.

1967
April
Secondary productivity in freshwater communities (PF). Prague, Czechoslovakia.

Methods for assessment of heat tolerance (HA). London, UK.

Methods for studying aphid populations (UM). Silwood Park, UK.

June
Project Telma: technical meeting on conservation of northern peatlands (CT/UNESCO/IUCN). Attingham Park, UK.

July
Methods for use in spider mite biological control project (UM). Sutton Bonington, UK.

September
Field methods for measurement of psychological performance (HA/CIBA Foundation). London, UK.

Working group on grasslands (PT). Saskatoon, Canada.

Exploration, utilization and conservation of plant gene resources (UM/FAO). Rome, Italy.

November
> Soil and seed inoculation (PP/UM/GIAM II). Addis Ababa, Ethiopia.
> Methods of study in soil ecology (PT/UNESCO). Paris, France.

1968
January
> Demography and sampling (HA). Utrecht, Netherlands.
> Biological control of rice stem borer (UM). Fukuoka, Japan.

March
> Inland water biology of South America (PF). Sante Fé, Argentina.
> Biological control of scale insects (UM). Riverside, California, USA.
> Conservation in the Western Mediterranean and the Sahara region (CT). Hammamet, Tunisia.

April
> Training course in bio-energetics (PT). Warsaw and Krakow, Poland.

May
> Inland water biology of tropical Africa (PF/UNESCO). Kampala, Uganda.
> Design and analysis in plankton sampling (PM in collaboration with IABO). Woods Hole, Mass., USA.

June
> Biology of man in Africa (HA). Warsaw, Poland.
> American Indian studies (HA/US IBP with PAHO). Washington, USA.

July
> Inland water biology of Southern Africa (PF). Grahamstown, South Africa.
> Marine food chains (PM). Aarhus, Denmark.

July–August
> Small mammals working group (PT). Oxford, UK.
> Productivity studies of temperate woodlands (PT). Gatlinburg, Tenn., USA.
> Marine mammals (PM). Cambridge, UK.

Appendix 4

August
Nitrogen fixation in relation to soil properties (PP). Adelaide, Australia.

Working group on data collection (CT). Monks Wood, UK.

August–September
Productivity in root systems and rhizosphere organisms (PT with USSR Academy of Sciences). Moscow and Leningrad, USSR.

September
Single cell protein (UM). Rutgers University, New Brunswick, N.J., USA.

Methods for the study of benthos (PM). Arcachon, France.

Evaluation of novel protein products (UM with IBP/Sweden and Wenner-Gren Foundation). Stockholm, Sweden.

Working meeting on analysis of ecosystems, tundra zone (PT). Ustaoset (Oslo), Norway.

Biological control of fruit flies (UM). Rome, Italy.

October
Physical performance capacity of adults (HA). Geneva, Switzerland.

November
Conservation of Pacific Oceanic Islands (CT/PM). Koror. Palau (Caroline Islands) and Guam.

December
Ecology and control of aquatic vegetation (PF/UNESCO). Paris, France.

1969
March–April
Training course in biostatistics (PT). Reading, UK.

May
South-east Asia regional seminar (PF/UNESCO). Kuala Lumpur and Malacca, Malaysia.

Check sheet working group (CT). Fenno-Scandinavia, Gotland, Sweden.

Microbial productivity in fresh waters (PF). Leningrad, USSR.

August

Symposium on the assessment of biological distance and affinity between human populations (HA). Utrecht, Netherlands.

September

Productivity of grassland ecosystems (PT). Saskatoon and Matador, Canada.

Models and methods for analysing photosynthetic systems (PP). Třebon, Czechoslovakia.

Increasing productivity of photosynthetic systems (PP). Moscow, USSR.

October

Productivity of woodland ecosystems (PT). Brussels, Belgium.

Productivity and conservation in northern circumpolar lands (Tundra conference, PT/CT/IUCN). Edmonton, Canada.

Productivity of forest ecosystems of the world (PT/UNESCO). Brussels, Belgium.

November–December

Survey of undisturbed oceanic islands; conservation and IBP research in India (CT; part of IUCN technical meeting). New Delhi, India.

December

Prospects of single-cell protein research for production of food and fodder (UM/III GIAM/UNESCO). Bombay, India.

1970

April

Grasslands biome working group (PT). Surfer's Paradise, Brisbane, Australia.

May

Symposium and working meeting on preliminary results of fresh-water production studies (PF/UNESCO). Warsaw, Poland.

Biological control of scale insects working group (UM). Riverside, California, USA.

Symposium on man in the Arctic (HA). Kiel, BRD.

Continuous monitoring in biological oceanography (SCOR/IBP Working Group No. 29). La Jolla, California, USA.

Appendix 4

June

Working group on social insects (PT). Warsaw, Poland.

August

Small mammals working group (PT). Helsinki, Finland.

September

Workshop on genetic modification of rhizobia (PP). Prague, Czechoslovakia.

Production, ecology and hydrological implications of aquatic macrophytes (PF/PP/PT/UNESCO–IHD). Maliuc, Romania.

Granivorous birds working group (PT). The Hague, Netherlands.

Technical meeting on nitrogen fixation (PP). Wageningen, Netherlands.

Tundra biome working group (PT). Kevo, Finland.

Bioenergetics and tropical ecosystems, East African regional meeting (PT/UNESCO). Kampala, Uganda.

Inter-biome synthesis (PT). Rome, Italy. Desert and grassland biome working groups (PT). Rome, Italy. Ecological stations working group (CT). Rome, Italy.

Biological control of aphids working group (UM). Versailles, France.

October

Check sheet survey co-ordination, E. Europe (CT). Prague, Czechoslovakia.

Biological control of scale insects (OILB/UM). Rabat, Morocco.

November

Photosynthesis under unusual conditions (SCOR/IBP Working Group No. 24). Nanaimo, BC, Canada.

Progress in soil microbiological studies (PT/UNESCO). Paris, France.

Technical meeting on leaf protein extraction (UM). Coimbatore, India.

December

Phytoplankton methods (SCOR/IBP Working Group No. 33). Rhode Island, USA.

Editorial Committee for handbook on microbial production (PF/UNESCO). Kyoto, Japan.

1971

February

Tundra biome: invertebrates working group (PT). Leningrad, USSR.

March

Tundra biome: decomposition working group (PT). Grange-over-Sands, UK.

Tundra biome: primary production working group (PT). Glenamoy, Ireland.

March–April

The biology of the Indian Ocean (IBP/PM, IIOE, IOC). Kiel, BRD.

April

Symposium on human biology of populations undergoing environmental change (HA). Blantyre, Malawi.

June

International conference on the biology of whales (IBP/PM, NSF, Smithsonian Inst.). Virginia, USA.

Coniferous forest workshop (PT). Kratte Masugn, Sweden.

August

Symposium: nitrogen economy of plant communities (IBP/PP, FAO, 12th Pacific Science Congress). Canberra, Australia.

Symposium: exploration and conservation of plant genetic resources (UM, 12th Pacific Science Congress). Canberra, Australia.

Symposium: interaction between organisms in insect pest regulation (UM, 12th Pacific Science Congress). Canberra, Australia.

Symposium: man in the pacific (HA, 12th Pacific Science Congress). Canberra, Australia.

Symposium: problems of nature conservation in the Pacific (CT, 12th Pacific Science Congress). Canberra, Australia.

September

Tundra biome: microbiology working group (PT). Bergen, Norway.

Working group: fruit flies (UM). Wadenswil, Switzerland.

Working group: aerobiology (UM). Exeter, UK.

October

Working conference (PM). Rome, Italy.

Working group: marine mammals (PM). Rome, Italy.

November
 Symposium: behaviour of ungulates in relation to management (PT).
 Calgary, Canada.

December–January
 Regional meeting: grassland biome (PT). Lamto, Côte d'Ivoire.

1972
January
 Working group: shallow lake ecosystems (PF). Lunz, Austria.
 Working group: interbiome microbiology (PT). Calgary, Canada.

March
 Tundra biome: ecosystem modelling (PT). San Diego, USA.

May
 Symposium: detritus and its ecological role in aquatic ecosystems
 (PF). Pallanza, Italy.
 Working group: humic acids (PF). Amsterdam, Netherlands.

June
 Symposium: ecophysical bases of arid zone ecosystem productivity
 (PT). Leningrad and Dushanbe, USSR.
 Working group: wetland ecosystems (PT, PP, PF). Mikołajki, Poland.
 Workshop: soil fauna and soil processes (PT). Louvain, Belgium.
 Symposium: modern methods in the study of microbial ecology (PT).
 Uppsala, Sweden.
 Working group: decomposition and soil processes (PT). Uppsala,
 Sweden.
 Symposium: breeding and productivity of barley (PP-P). Kroměříž,
 Czechoslovakia.

July
 Aerobiology: international working group (UM). Boulder, USA.

August
 Symposium: biology of the seal (PM). Guelph, Canada.
 Woodland biome workshop (PT). Oak Ridge, USA.
 Grassland–Tundra biome workshop (PT). Fort Collins, USA.
 Social insects: review and synthesis. Aarhus, Denmark. Workshop:
 integrated pest management (UM, 14th Congress on Entomology).
 Canberra, Australia.

August–September
Section working group (HA). Seattle, USA.

September
Novel protein sources (UM, 11th International Congress of IUNS). Mexico City, Mexico.
Desert biome workshop (PT). Logan, USA.
Symposium: results of PF projects (PF). Reading, UK.
Symposium: the biological effects of inter-ocean canals (PM). Monaco.

October
Working group: synthesis of CT projects (CT). London, UK.

November
Symposium: legume inoculation – science and technology (PP-N). New Delhi, India.

1973

March
Exploration, utilization and conservation of plant genetic resources (UM). Rome, Italy.
Grassland synthesis, Chief Editors' meeting (PT). Paris, France.

March–April
Tundra working groups: synthesis (PT). Dublin, Ireland.

April
Photosynthetic systems: their functioning in different environments (PP). Aberystwyth, UK.

May
Freshwater productivity: synthesis and editorial (PF). Leningrad, USSR.
Wetlands working group (PP/PF/PT). Warsaw, Poland.

June
Zoobenthos editorial and synthesis (PF). Stirling, UK.
Primary production editorial and synthesis (PF). Windermere, UK.

June–July
Circumpolar peoples: synthesis meeting (HA). Rekjavik, Iceland.

July
Regional meeting on semi-natural temperate grasslands and synthesis (PT). Warsaw, Poland.
Freshwater fish: editorial and synthesis (PF). Reading, UK.

Appendix 4

August
   Microbiology and decomposition in tundra ecosystems (PT). Fairbanks and Pt Barrow, USA.
   Editorial and synthesis (HA). Detroit, USA.

September
   Review and synthesis meeting: granivorous birds (PT). Warsaw, Poland.
   Biological control synthesis (UM). Berkeley, USA.
   Eastern Mediterranean (PM). Malta.
   Nitrogen fixation and the biosphere (PP). Edinburgh, UK.
   Zooplankton: editorial and synthesis (PF). Stockholm, Sweden.

September–October
   Woodlands synthesis (PT). Göttingen, BDR.

October
   Zoobenthos: editorial and synthesis (PF). Lunz, Austria.
   CT editorial (CT). London, UK.

November
   Synthesis of marine mussels (PM). London, UK.

December
   Editorial and synthesis (CT). London, UK.

1974
January
   Regional meeting on tropical grasslands and grassland synthesis (PT). Varanasi, India.
   Wetlands: synthesis (PF/PP/PT). Třeboň, Czechoslovakia.

March
   Woodlands decomposition (PT). Merlewood, UK.
   IBP Chief Editors of synthesis volumes. London, UK.
   PT Chief Editors of synthesis volumes (PT). London, UK.

April
   Synthesis meeting: tundra (PT). Abisko, Sweden.

April–May
   Editorial and synthesis (PF). Pallanza, Italy.

June
   Grey mullets and their culture (PM). Haifa, Israel.

178

August
Decomposition and soil processes synthesis (PT). Louvain, Belgium.

September
Woodlands synthesis (PT). The Hague, Netherlands.
IBP intersectional symposia. The Hague, Netherlands.
Granivorous birds synthesis. (PT) Ft Collins, USA.

4b. General Assemblies, SCIBP and Bureau Meetings

	General Assembly	SCIBP Meetings	Meeting of Bureau and Bureau Conveners
1964	I GA, Paris, July	I SCIBP, July, Paris	
1965		II SCIBP, February, Rome	I Bureau, October
			II & III Bureau, February, Rome
			IV Bureau, September, London
1966	II GA Paris, April	III SCIBP, Paris, April	V Bureau, April, Paris
			VI Bureau, October, London
1967		IV SCIBP, Paris, March	VII Bureau, March, Paris
			VIII Bureau and Conveners October, London
1968	III GA, Varna, April	V SCIBP, Varna, April	
1969			IX Bureau and Conveners, March/April, Berne
		VI SCIBP, September/October, London	
1970			X Bureau and Conveners, April, Amsterdam
	IV GA Rome, September/October	VII SCIBP, Rome, October	
1971		Part SCIBP and National Committees, Canberra, September	XI Bureau and Conveners, June, Budapest
1972			XII Bureau and Conveners, March, London
	V GA, Seattle, September	VIII SCIBP, Seattle, September	
1973			XIII Bureau and Conveners, April, Paris
1974		IX SCIBP, London, March	

5. Publications

5a. Publications by SCIBP

IBP News

1 November 1964. Record of 1st General Assembly and SCIBP, Paris.
2 February 1965. Sectional programmes for Phase I.
3 June 1965. IInd Meeting of SCIBP, Rome.
4 November 1965. Sectional committee meetings.
5 March 1966. Record of IVth Meeting of SCIBP Bureau, sectional reports.
6 June 1966. Record of IInd General Assembly, Paris.
7 October 1966. Reports on sectional activity.
8 May 1967. Record of IVth Meeting of SCIBP, Paris.
9 June 1967. Sectional programmes for Phase II.
10 February 1968. Sectional news.
11 March 1969. Record of IIIrd General Assembly, Varna, Bulgaria.
12 March 1969. Summary record: symposium on biological bases of productivity.
13 January 1969. Index of national projects: Section PT.
14 March 1969. Index of national projects: Section PP (Photosynthesis).
15 March 1969. Index of national projects: Section PP (Nitrogen fixation).
16 February 1969. Index of national projects: Section CT.
17 March 1969. Index of national projects: Section PF.
18 March 1969. Index of national projects: Section PM.
19 February 1969. Index of national projects: Section HA.
20 March 1969. Index of national projects: Section UM.
21 September 1970. Record of VIth Meeting of SCIBP, London.
22 September 1970. Index of national projects: Section HA, Revised edition.
23 December 1972. Record of IVth General Assembly, Rome 1970 and Record of Vth General Assembly, Seattle, USA 1972.
24 March 1973. Programme for Phase III, Synthesis and Transfer.
25 June 1974. Record of IXth and final meeting of SCIBP, London.

The Biosphere

Bulletin of the International Biological Programme. 10 issues from 1967 to 1971.

Appendix 5

IBP Review

The International Biological Programme: a statement from the Special Committee for the IBP, August 1968 to the '*Intergovernmental Conference of Experts on the Scientific Basis for Rational Use and Conservation of the Resources of the Biosphere*', 28 pp. UNESCO, Paris (September 1968).

International Biological Programme 1969 Review, 48 pp. SCIBP, London (September 1969).

International Biological Programme 1970 Review, 52 pp. SCIBP, London (September 1970).

International Biological Programme 1972 Review, 65 pp. SCIBP, London (August 1972).

IBP Directory

Directory of National Participation in IBP, 32 pp. SCIBP, London (September 1969).

Directory of National Participation in IBP, revised edit., 31 pp. SCIBP, London (July 1972).

IBP Handbooks

Published for SCIBP by Blackwell Scientific Publications Ltd., Oxford.

1 *A Guide to the Human Adaptability Proposals*, 2nd edition. Ed. J. S. Weiner (1969).
2 *Methods for Estimating the Primary Production of Forests*. Ed. P. J. Newbould (1967) (reprint 1970).
3 *Methods for Assessment of Fish Production in Fresh Waters*. Ed. W. E. Ricker (1968; 2nd edition 1971).
4 *Guide to the Check Sheet for IBP Areas*. Ed. G. F. Peterken (1968; reprinted 1970).
5 *Handbook to the Conservation Section of the IBP*. Ed. E. M. Nicholson (1968).
6 *Methods for the Measurement of the Primary Production of Grassland*. Eds. C. Milner & R. E. Hughes (1968).
7 *A Practical Guide to the Study of the Productivity of Large Herbivores*. Eds. F. B. Golley & H. K. Buechner (1969).
8 *Methods for Chemical Analysis of Fresh Waters*. Ed. H. L. Golterman, asst. R. S. Clymo (1969) (reprint 1970).

9 *Human Biology: A Guide to Field Methods.* Eds. J. S. Weiner & J. A. Lourie (1969).
10 *Methods for the Measurement of Psychological Performance.* Ed. S. Biesheuvel (1969).
11 *Genetic Resources in Plants – Their Exploration & Conservation.* Eds. O. H. Frankel & E. Bennett (1970).
12 *A Manual on Methods for Measuring Primary Production in Aquatic Environments.* Ed. R. Vollenweider (1969) (revised 1974).
13 *Productivity of Terrestrial Animals – Principles and Methods.* Eds. K. Petrusewicz & A. Macfadyen (1970).
14 *A Handbook of Field Methods for Research on Rice Stem-Borers & their Natural Enemies.* Eds. T. Nishida & T. Torii (1970).
15 *A Handbook for the Practical Study of Root-Nodule Bacteria.* Ed. J. M. Vincent (1970).
16 *Methods for Studies of Marine Benthos.* Eds. N. A. Holme & A. McIntyre (1971)
17 *Methods for Estimation of Secondary Productivity in Fresh Waters.* Ed. W. T. Edmonson (1971).
18 *Methods of Study in Quantitative Soil Ecology: Population, Production & Energy Flow.* Ed. J. Phillipson (1971).
19 *Methods of Study of Ecology of Soil Micro-organisms.* Eds. D. Parkinson, T. Gray & S. Williams (1971).
20 *Leaf Protein.* Ed. N. W. Pirie (1971).
21 *Project Aqua List.* Eds. H. Luther & J. Rzóska (1971).
22 *Instruments for Micro-Meteorology.* Ed. J. L. Monteith (1973).
23 *Techniques for the Assessment of Microbial Production and Decomposition in Fresh Waters.* Eds. Y. Sorokin & H. Kadota (1972).
24 *Methods for Studies of Ecological Bio-Energetics.* Eds. R. Z. Klekowski, W. Grodziński & A. Duncan (1975).

Other publications

An Approach to the Rapid Description and Mapping of Biological Habitats. M. E. D. Poore & V. C. Robertson. IBP Central Office (1964).

Analysis of Ecosystems: Tundra Biome. Proceedings of an IBP/PT Working Meeting held at Kevo, Finland, in September 1970. Ed. O. W. Heal. IBP Central Office (1971).

IBP/CT Progress Report, 1971. E. M. Nicholson & G. L. Douglas. IBP Central Office (1971).

Appendix 5

Progress in Growth and Physique Studies. Eds. P. B. Eveleth & J. M. Tanner. IBP Central Office (1971).

Check List of Pacific Oceanic Islands. Gina Douglas. Reprinted from *Micronesica,* 5 (2), 1969. IBP Central Office (1971).

Human Biology of Environmental Change. Proceedings of a Conference held at Blantyre, Malawi, April, 1971. Ed. D. J. M. Vorster. IBP Central Office (1972).

IBP World Catalogue of Rhizobium Collections. Compiled by O. N. Allen & Eva Hamatová, ed. F. A. Skinner. IBP Central Office (1973).

5b. International volumes published elsewhere

1965

Primary Productivity in Aquatic Environments. Proceedings of an IBP/PF Symposium at Pallanza. Ed. C. R. Goldman. 457 pp. Memorie dell'- Istituto Italiano di Idrobiologia (1965).

1966

Human Adaptability and its Methodology. Proceedings of a Symposium sponsored by IUPS and IBP, Kyoto, 12–44 September 1965. Ed. H. Yoshimura & J. S. Weiner. Japan Society for the Promotion of Sciences, Tokyo (1966).

The Biology of Human Adaptability. Ed. Paul T. Baker & J. S. Weiner. Clarendon Press, Oxford (1966).

Man-Made Lakes, Institute of Biology Symposia No. 15. Ed. R. H. Lowe-McConnell. Academic Press, London (1966).

Human Adaptability to Environments and Physical Fitness. Proceedings of a Symposium held at New Delhi, September 1965. Ed. M. S. Malhotra. Defence Institute of Physiology and Allied Sciences, Madras 3 (1966).

Functioning of Terrestrial Ecosystems at the Primary Production Level. Proceedings of a Symposium in July 1965 at Copenhagen organised by UNESCO in association with PT and PP. Ed. F. Eckardt. UNESCO SC.NS.66/XII.5/AF (1966).

1967

Chemical Environment in the Aquatic Habitat: Proceedings of an IBP Symposium held in Amsterdam and Nieuwersluis, 10–16 October 1966. Ed. H. L. Golterman & R. S. Clymo. Koninklijke Nederlandse Akademie van Wetenschappen (1967).

184

Secondary Productivity of Terrestrial Ecosystems: Principles and Methods, 2 vols. Proceedings of Working Meeting held in Jablonna, 31 August to 6 September 1966. Ed. K. Petrusewicz. Polish Academy of Sciences Institute of Ecology (1967).

The Biological Basis of Freshwater Fish Production. Proceedings of an IBP/PF meeting at Reading. Ed. S. D. Gerking. Blackwell Scientific Publications, Oxford (1967).

Photosynthesis of Productive Systems. Ed. A. A. Nichiporovich, translated from the Russian by N. Kaner and edited by J. L. Monteith. Academy of Sciences of the USSR. Israel Program for Scientific Translations, Jerusalem (1967).

1968

Problems in Human Adaptability: Proceedings of a meeting held in Warsaw, 26–30 April 1965. Polish Academy of Sciences in Materialy I Prace Antropologiczne Nr. 75, Wroclaw (1968).

Primera Reunion Regional de Limnologia Latinoamericana, 14–18 Marzo 1968. Resoluciones y Recomendaciones. Informes (with English summaries). Consejo Nacional (1968).

1969

Methods of Productivity Studies in Root Systems and Rhizosphere Organisms. Proceedings of an International Symposium held at Moscow, 28 August to 12 September 1968. USSR Academy of Sciences. Reprinted by SCIBP (1969).

Energy Flow through Small Mammal Populations. Proceedings of an IBP/PT meeting held in Oxford, 29 July–2 August 1968. Ed. K. Petrusewicz & L. Ryszkowski. Polish Scientific Publishers, Warsaw (1969/70).

Malawi Science Conference. Proceedings of Conference held at Chancellor College, University of Malawi, 25 and 26 July 1969.

SE Asian Regional Meeting: Biology of Inland Waters. Proceedings of a meeting held at Kuala Lumpur and Malacca, 5–11 May 1969. UNESCO Regional Office, Djakarta (1969).

Conservation of Pacific Islands. Proceedings of Technical Meeting held at Koror, Palau Island, and Guam, 1968. *Micronesica*, **5** (1969).

Report of the Regional Meeting of Hydrobiologists in Tropical Africa. 20–28 May 1968. Makerere, Uganda. UNESCO Regional Centre for Science and Technology, Nairobi (1969).

Appendix 5

1970

Trace Element Metabolism in Animals. Proceedings of WAAP/IBP International Symposium, Aberdeen, Scotland, July 1969, 549 pp. Eds. C. F. Mills. E. & S. Livingstone, Edinburgh & London (1970).

Analysis of Temperate Forest Ecosystems. Based on an IBP/PT meeting held in Tennessee, USA, from 28 July–3 August 1968. Ed. D. E. Reichle, Springer-Verlag, Heidelberg & New York (1970).

Methods of Study in Soil Ecology: Proceedings of the Paris Symposium, Méthodes d'études de l'écologie du sol: Actes du colloque de Paris. Proceedings of a joint UNESCO/IBP-PT meeting held in Paris from 7–10 November 1967. Ed. J. Phillipson. UNESCO, Paris (1970).

Biology of Man in Africa: Proceedings of a technical meeting held in Warsaw, 24–7 June 1968. Polish Academy of Sciences in Materialy I Prace Antropologiczne Nr. 78, Wroclaw (1970).

Productivity Problems of Freshwaters. Preliminary papers for a UNESCO/IBP-PF symposium held at Kazimierz Dolny, Poland, 6–12 May 1970, 2 vols. Polish Academy of Sciences (1970).

Prediction and Measurement of Photosynthetic Productivity. Proceedings of an IBP/PP Technical Meeting held at Třeboň, Czechoslovakia, 14–21 September 1969. Centre for Agricultural Publishing and Documentation, Wageningen (1970).

Soil and Seed Inoculation. Proceedings of an IBP/PP Technical Meeting held at Addis Ababa, 5–12 November 1967. *Plant and Soil*, **32**, 543–736 (1970).

Bibliography on Methods of Studying the Marine Benthos, IBP/PM. Ed. A. D. McIntyre, FAO Fisheries Technical Paper No. 98 (1970).

Productivity and Conservation in Northern Circumpolar Lands. Proceedings of an IUCN/IBP Conference held in Edmonton, Canada from 15–17 October 1969. Ed. W. A. Fuller & P. G. Kevan. IUCN Publications New Series No. 16 (1970).

Grassland Ecosystems: Reviews of Research. Proceedings of the second meeting of the IBP/PT Grasslands Working Group, held in Saskatoon and Matador, Canada, from 5–10 September 1969. Ed. R. T. Coupland & G. M. Van Dyne. Range Science Dept., Science Series No. 7, Colorado State University (1970).

Progress in Soil Microbiological Studies. Based on a UNESCO/IBP-PT meeting held in Paris from 17–19 November 1970. Ed. D. Parkinson. University of Calgary (1970).

186

Evaluation of Novel Protein Products. Proceedings of a Symposium, Stockholm, Sweden, September 1968. Eds. A. E. Bender, R. Kihlberg, B. Löfqvist & L. Munck. Pergamon Press, London (1970).

1971

Symbols, Units and Conversion Factors in Studies of Fresh Water Productivity. Ed. G. G. Winberg (Leningrad) and collaborators. IBP Central Office (1971).

Proceedings of the international symposium on the biological productivity of aquatic and swamp macrophytes, held at Maliuc, Romania, 1–10 September 1970. *Hidrobiologia,* **12,** 5–408 (1971).

Productivity of Forest Ecosystems. Proceedings of the Brussels Symposium. Ecology and Conservation No. 4 UNESCO (1971).

Global Impacts of Applied Microbiology. IBP/UNESCO Symposium on Single Cell Protein (SCP), 7–12 December 1969, pp. 403–594. University of Bombay, India (1971).

Plant Photosynthetic Production: Manual of Methods. Eds. Z. Sesták, J. Catsky & P. G. Jarvis. Junk, The Hague (1971).

Integrated Experimental Ecology. Methods and Results of Ecosystem Research in the German Solling Project. Ed. H. Ellenberg. Springer-Verlag, Berlin, Heidelberg & New York, Ecological Studies No. 2 (1971).

Secondary Productivity in Small Mammal Populations. Proceedings of an IBP/PT meeting held in Helsinki, Finland, from 24–8 August 1970. Reprinted from *Annales Zoologici Fennici,* **8,** 1–185 (1971).

Systems Analysis in Northern Coniferous Forests. IBP Workshop, Kratte Masugn, Sweden, 14–18 June 1971. Ed. Th. Rosswall, Swedish Natural Science Research Council. Bulletin from the Ecological Research Committee No. 14 (1971).

1972

Ecophysiological Foundation of Ecosystem Productivity in Arid Zones. International Symposium of IBP/PT, USSR, 7–19 June 1972, USSR Academy of Sciences (1972).

Preliminary Report on the Results of the International Rice Adaptation Experiments, 1970. Eds. E. J. A. Khan *et al.* IRAE No. 3, IBP Central Office (1972).

Appendix 5

Tundra Biome. Proceedings of the IVth International meeting on biological productivity of the Tundra, held at Leningrad, USSR, in October 1971. Eds. F. E. Wielgolaski & Th. Rosswall (1972).

Proceedings of Symposium on Rice Stem Borers, Related Pests and their natural enemies. 12th Pacific Science Congress Symposium No. 9A. 1 *Mushi,* **43,** suppl. 59 pp. (1972).

Biological Nitrogen Fixation in Natural and Agricultural Habitats. Proceedings of the Technical Meeting on Biological Nitrogen Fixation, IBP/PP, Prague & Wageningen, September 1970. *Plant and Soil* (1972).

Proceedings of the Symposium on the Nitrogen Economy of Plant Communities, held in Canberra, Australia at the 12th Pacific Science Association, 18–19 August 1971. Ed. F. J. Bergerson. *Soil Biology and Biochemistry* (1972).

Aphid Technology. Ed. H. F. van Emden. Academic Press, London & New York (1972).

Productivity, Population Dynamics and Systematics of Granivorous Birds. Proceedings of a Working Group held at The Hague, September 1970. Eds. S. C. Kendeigh & J. Pinowski. Polish Scientific Publishers, Warsaw (1972).

Productivity Investigations on Social Insects and their role in the ecosystems. Proceedings of Working Group. Warsaw, Poland, June 1970. Eds. M. V. Brian & J. Petal. *Ekologia Polska,* **20,** Nos. 1–19 (1972).

Detritus and its role in aquatic ecosystems. Proceedings of an IBP/ UNESCO Symposium, Pallanza, Italy, May 1972. Ed. U. Melchiorri-Santolini & J. W. Hopton. *Memorie dell'Istituto Italiano di Idrobiologia,* **29** suppl. (1972).

Ecology of Water Weeds in the Neotropics. D. S. Mitchell & P. A. Thomas. UNESCO Technical Papers in Hydrology, No. 12. A contribution to the International Hydrological Decade. UNESCO, Paris (1972).

Modeling Forest Ecosystems, Report of International Woodlands Workshop, Oak Ridge, USA, August 1972. Oak Ridge National Laboratory EDFB-IBP-73-7 UC-48 (1973).

1973

Modern Methods in the study of Microbial Ecology. Proceedings of a Symposium held at Uppsala, Sweden, June 1972. Ed. Th. Rosswall. Swedish Natural Research Council Bulletin from the Ecological Research Committee, No. 17 (1973).

188

Proceedings of the IBP Wetlands Symposium, 11–18 June 1972. Ed. A. Szczepánski. *Polskie Archiwum Hydrobiologii,* **20** (1) (1973).

Monitoring Life in the Ocean. Report of Working Group 29 on Monitoring in Biological Oceanography. SCOR-ACMRR-UNESCO-IBP/PM (1973).

Nature Conservation in the Pacific. Proceedings of a Symposium at the 12th Pacific Science Congress, August 1971. Eds. A. B. Costin & R. H. Groves. Australian National University Press (1973).

The Biology of the Indian Ocean. Proceedings of a Symposium at Kiel, FRD, March/April 1971. Ed. B. Zeitzschel. Springer-Verlag, Berlin, Heidelberg & New York. Ecological Studies No. 3 (1973).

Regional Symposium on Conservation of Nature – Reefs and Lagoons. Proceedings and papers of a Symposium held at Noumea, New Caledonia, August 1971. Ed. H. F. I. Elliott. South Pacific Commission (1973).

Primary Production and Production Processes, Tundra Biome. Proceedings of a Conference, Dublin, 1973. Eds. L. C. Bliss & F. E. Wielgolaski. Edmonton (1973).

1974

Proceedings of a Symposium and synthesis meeting on Tropical Grassland Biome, Varanasi, India, January 1974. Banares Hindu University, Varanasi, India (1974).

Aquatic Vegetation and its Use and Control. Ed. D. S. Mitchell. A contribution to the International Hydrobiological Decade. UNESCO, Paris (1974).

The behaviour of ungulates and its relation to management (2 vols.). Papers of an international Symposium held at Calgary, November 1971. Eds. V. Geist & H. F. I. Elliott. IUCN Publications, No. 24 (1974).

Soil Organisms and Decomposition in Tundra. Proceedings of a synthesis meeting, Fairbanks, 1973. Eds. A. J. Holding *et al.* Stockholm (1974).

The Whale Problem. A status report resulting from the Conference on the Biology of Whales, Virginia, USA, 1971. Ed. W. E. Scherill. Harvard University Press (1974).

In press

Biology of the Seal. Proceedings of a Symposium, Guelph, Canada, 1972.

Bio-Social Inter-relations in Population adaptations. Proceedings of a Symposium at Detroit, USA, August 1972. Eds. F. Johnston & E. Watts.

Appendix 5

Productivity of World Ecosystems. Proceedings of a Symposium, Seattle, Washington, 1972. National Academy of Sciences, 1975.

5c. National programmes, reports and synthesis volumes

ARGENTINA

IBP Argentine Report 1970–1972 (in Spanish and English), National Committee for Scientific and Technical Research, Buenos Aires.

AUSTRALIA

IBP Provisional Australian Programme, Australian Academy of Sciences, Canberra 1966.

IBP Australian Programme, Australian Academy of Sciences, Canberra, 1969.

IBP Australian Progress Report, 1967–1970, Australian Academy of Sciences, Canberra 1970.

National synthesis include *Conservation of major plant comunities in Australia and Papua New Guinea.* Eds. R. L. Specht *et al.* CSIRO (1974); and reports of HA New Guinea project published jointly with UK.

AUSTRIA

Preliminary Framework of the Austrian Programme for the IBP, February 1966, Vienna (roneo).

Additional Projects for the Framework of the Austrian Programme for the IBP, December 1966, Vienna (roneo).

Progress Report 1968 of the Austrian National Committee for the IBP, February 1969.

Progress Report 1969 of the Austrian National Committee, May 1970 (roneo, IBP).

BELGIUM

Programme National Belge pour le Programme Biologique International, October 1966 (roneo).

BRAZIL

Brazilian Contribution for the International Biological Programme (Preliminary Projects) (in English and Portuguese), Conselho Nacional de Pesquisas, 1968.

190

BULGARIA

Proposed Researches of the Bulgarian National Committee, April 1966.
Progress Report 1968 of the Bulgarian National Committee for the IBP, Sofia, 1968.

CANADA

Provisional Canadian Contribution to the IBP, April 1966.
Canadian Contribution to the IBP, revised December 1966.
Canadian Participation in the IBP, Report no. 1, May 1967.
Canadian Report no. 2, January 1969.
Summary Annual Reports from Projects for 1969, CCIBP, May 1970.
Organization and activities 1964–1970, CCIBP, May 1971.
Summary Annual Reports from Projects for 1970. CCIBP, June 1971.
Summary Annual Reports from Projects for 1971. CCIBP, June 1972.
Terminal Summary and Status of Projects, CCIBP, June 1973.
National syntheses include a semi-popular summarisation of results and *Productivity of High Arctic Tundra*, both now in preparation. Other syntheses are under discussion.

CZECHOSLOVAKIA

Provisional Czechoslovak Contribution to Phase One of the IBP, Czechoslovak Academy of Sciences, Prague, 1966.
The Czechoslovak Contribution to the International Biological Programme, Czechoslovak Academy of Sciences, Prague, 1968.
Progress Report 1968, 1969 of the Czechoslovak National Committee of the International Biological Programme (roneo).
Summary Report on National Results of IBP (in preparation).

DENMARK

Preliminary Programme for the Participation of Denmark in IBP, February 1967.
(See also Scandinavia)

EAST AFRICA – TANZANIA

International Biological Programme: Tanzanian Contribution, February 1969.

EAST AFRICA – UGANDA

National Programme for the IBP, April 1969.

Appendix 5

ECUADOR
Ecuadorian Nature and Biota Research Programme, 1968 (roneo, IBP, SCIBP 39/68).
Contribuciones del Ecuador al Programa Internacional de Biologia (Bol. no. 2 del IECN-IBP), March 1969 (In Spanish).
Contributions of Ecuador to the IBP, September 1970.

FINLAND
Preliminary Programme for the Participation of Finland in IBP, December 1966 (roneo).
(See also Scandinavia)

FRANCE
Projet de Contribution Française au Programme Biologique International, Comité Français, March 1966 (roneo).
Projet de Contribution Française au Programme Biologique International, Comité Français, February 1967 (roneo).
Compte-rendu d'Activité de la Participation Française pour l'année 1968, Comité Français, January 1969 (In French).
Compte-rendu d'Activité de la Participation Française pour l'année 1969, Comité Français, January 1970 (In French).
Compte-rendu d'Activité de la Participation Française, Comité Francais, June 1972 (In French).

GERMANY (DEMOCRATIC REPUBLIC OF)
National-Programm der DDR für das Internationale Biologische Programm, March 1966.

GERMANY (FEDERAL REPUBLIC OF)
Contributions of the Federal Republic of Germany.
German Research Association, Bad Godesberg, May 1967. (In English and German).

GHANA
Draft programme for the International Biological Programme of Ghana (Sections PT/CT/UM), typescript, May 1968.

HUNGARY
Hungarian Contribution to the IBP, Hungarian Academy of Sciences, June 1968.
Final Report in process of preparation.

192

INDIA

IBP Indian Programme, March 1967.

Indian National Programme, Indian National Committee for Biological Sciences, August 1967.

Indian Participation in the International Biological Programme: Progress Report 1966–68 National Institute of Sciences of India, Delhi, August 1969.

Indian Participation in the International Biological Programme: Progress Report 1969–71. Indian National Science Academy, New Delhi, August 1972.

INDONESIA

Preliminary Communication on the Indonesian National Biological Programme, 1966 (roneo).

Provisional Indonesian Programme, Council for Sciences of Indonesia, Djakarta, January 1967 (roneo).

IRELAND (REPUBLIC OF)

Irish Contribution: Interim Report 1971, Royal Irish Academy, Dublin, June 1971.

ISRAEL

Provisional Israeli Programme, March 1966 (roneo).

Provisional Israeli Contribution to the IBP, The Israel Academy for Sciences and Humanities, Jerusalem, March 1967.

ITALY

Proposed Researches of Italian National Committee for IBP, February 1966.

Italian Revised Programme, December 1966 (roneo).

Progress Report 1969 of the Italian National Committee of IBP and List of Biotopes. Consiglio Nazionale delle Richerche, Roma, *Quaderni de 'La Ricerca Scientifica' no. 65, April* 1970.

JAPAN

Proposed Researches of Japanese National Committee for IBP, December 1964 (roneo).

Report on the Activity and Future Plan of Japanese National Committee for IBP, February 1966 (roneo).

Interim Report of the Japanese National Committee for IBP, Its Activity in 1966 and Research Projects Proposed for Phase II.

The Science Council of Japan, Tokyo, February 1967.

Progress Report 1967 of the Japanese National Committee for IBP, Report no. 4, The Science Council of Japan, Tokyo, February 1968.

Progress Report for 1968 of the Japanese National Committee for IBP. Report no. 5, The Science Council of Japan, Tokyo, February 1969.

Progress Report 1969 of the Japanese National Committee for IBP. Report no. 6, The Science Council of Japan, Tokyo, March 1970.

Japanese contribution to IBP in 1970, Report no. 7, The Science Council of Japan, Tokyo, March 1971.

Japanese contribution to IBP in 1971, Report no. 8, The Science Council of Japan, Tokyo, March 1972.

Japanese contribution to IBP in 1972, Report no. 9, The Science Council of Japan, Tokyo, August 1973.

National syntheses to be published from 1974–75 include an overall summary of the Japanese contribution, 5 volumes on PT ecosystem studies, grasslands, forests and secondary productivity, 3 volumes on PP crop photosynthetic production, ecophysiology and nitrogen fixation, 1 volume on CT, 1 volume on PF inland waters, 1 volume on PM productivity of coastal biocenoses, 3 volumes on HA and 3 volumes on UM gene pools, plant adaptations and biological control.

KOREA (REPUBLIC OF)

Research Project of the Korean National Committee for the IBP, Korean National Academy of Sciences, Seoul, February 1966.

Report for the IBP no. 1 Korean National of Sciences, Seoul, December 1967.

Report for the IBP no. 2 Korean National Academy of Sciences, Seoul, December 1968 (English and Korean).

Report for the IBP no. 3, Korean National Academy of Sciences, Seoul, December 1969 (English and Korean).

Report for the IBP no. 4, Korean National Academy of Sciences, Seoul, December 1970 (English and Korean).

Report for the IBP no. 5, Korean National Academy of Sciences, Seoul, May, 1971 (English and Korean).

A brief report on activity of Korean IBP Project to SCIBP meeting. Yung Sun Kang, Chairman, August 1971 (typescript).

Report for the IBP no. 6, Korean National Academy of Sciences, Seoul, May 1972 (English and Korean)

MALAWI

Report on the Initiation in Malawi of the International Biological Programme, February 1969.

MALAYSIA

Progress Report for the Malaysian National Committee for IBP, 1968, August 1969 (roneo).
Overall synthesis and Review of Results of Projects at Pasoh and Tasek in preparation.

NETHERLANDS

Communication to SCIBP from the Netherlands National Committee for the IBP, March 1966 (roneo).
Dutch IBP Projects, Royal Netherlands Academy of Sciences and Letters, Biological Council, Amsterdam, February 1967 (roneo).
Progress Report, 1966–7, Netherlands National Committee for IBP, Amsterdam, December 1968.
Progress Report, 1968–9, Netherlands National Committee for IBP, Amsterdam, September 1970.
Final Report, 1966–71, Netherlands National Committee for IBP, Amsterdam, 1974.

NEW ZEALAND

IBP in New Zealand: Report on Progress to Date, March 1967.
New Zealand National Programme, Royal Society of New Zealand, March 1968.

NIGERIA

Biological Research Projects approved by the National Committee for IBP in Nigeria (1965–6) (roneo).
Nigerian Participation in the IBP, Progress Report (1970). Nigerian National Committee, Lagos.

NORWAY

Norwegian Programme, Norwegian National Committee for IBP, January 1967 (roneo).
IBP i Norge, Arsrapport, 1969.
IBP in Norway: Methods and Field Data (in press, about 1000 pp).
(See also Scandinavia)

Appendix 5

PHILIPPINES
Programme of the Philippines National Committee, March 1966 (roneo).
Philippine Programme for the International Biological Programme, Bulletin no. 49 of the National Research Council of the Philippines, Quezon City, December, 1968.

POLAND
Proposed Researches of the Polish National Committee for the IBP, 1966.
Polish Contribution to Definite Phase of the IBP: Programme of Investigations, Polish Academy of Sciences, Warsaw, March 1967.
Contribution to definite phase of the International Biological Programme (programme of investigations), revised edition, Polish Academy of Sciences, Warsaw, 1968.
Progress Report 1967–68 of the Polish National Committee for International Biological Programme, Polish Academy of Sciences, Warsaw, 1969.
National synthesis volume in press.

RHODESIA
Rhodesian Contribution to the International Biological Programme. Scientific Council of Rhodesia, Salisbury, 1970.

ROMANIA
Research Work Plan 1968–72, April 1968 (roneo).
Contribution of Romania to the IBP for 1968 and 1969 (Summary, UM Section). Academy of Sciences of the Socialist Republic of Romania, Bucharest, 1970.

SCANDINAVIA
Scandinavian Contributions to International Biological Programme, March 1968 (joint publication of four-nation regional committee).
Progress Report 1968: Scandinavian countries Denmark, Finland, Norway, Sweden, August 1969.
Progress Report 1969–70; of Scandinavian countries Denmark, Finland, Norway, Sweden, 1971.
Progress Report 1971: Scandinavian countries Denmark, Finland, Norway, Sweden, May 1972.
Progress Report 1972: Scandinavian countries Denmark, Finland, Norway, Sweden, October 1973.

196

Final Report Scandinavian IBP Committees (in press).
Fennoscandian Tundra Ecosystems, vols. 1 & 2. Ed. F. E. Wielgolaski (in press: Springer-Verlag).
Fennoscandian HA project in Inari on Sholt Lapps: synthesis planned.

SOUTH AFRICA
Proposed South African Programme for Participation with IBP, March 1966 (roneo).
South African Participation in the IBP, March 1967 (roneo).
South African Participation in the International Biological Programme, South African Council for Scientific and Industrial Research, June 1969.
Summary Report to SCIBP, South African Participation in the IBP, Science Co-operation Division, CSIR, Pretoria, April 1970.
Summary Report 1967–1972 of the South African National Committee for IBP, August 1972.

SPAIN
Draft Programme of the Spanish National Committee of the IBP, April 1966.
Programme of the Spanish Contribution to the IBP, Madrid, 1966.

SWEDEN
Swedish IBP Programme, April 1966.
BIORED (Biological Resources Development Teams), Parts I & II.
The Swedish Contribution to the IBP, February 1967 (roneo).
(See also Scandinavia).
Swedish Final Report (in preparation).
Swedish Tundra Project (to be published in 1975).

THAILAND
Report on National Programme in Terrestrial Productivity: Study of Productivity in Tropical Forests, February 1969 (roneo, SCIBP 8/69).

UNITED KINGDOM
Provisional United Kingdom Contribution to Phase One of the IBP, The Royal Society, London, August 1965.
The United Kingdom Contribution to the IBP, The Royal Society, London, February 1967.
The United Kingdom Contribution to the IBP, 1970. Progress Report. The Royal Society, London, September 1970.

The United Kingdom Contribution to the IBP, 1972. Progress Report. The Royal Society, London, 1972.

Semipopular summarisation of results (to be published in 1975).

PT results of woodlands, grasslands and tundra projects to be published by Springer-Verlag in 1975.

PF Lake George results and HA New Guinea results published in *Proceedings of The Royal Society.*

PF Loch Leven results published by the Royal Society of Edinburgh.

HA Israel Studies published in *Philosophical Transactions of the Royal Society.*

HA New Guinea Studies published in *Philosophical Transactions of the Royal Society.*

UNITED STATES OF AMERICA

Preliminary Framework of the US Programme of the IBP, Publication no. 1, National Academy of Sciences, National Research Council, Washington DC, August 1965.

US Participation in the IBP. Report no. 2. National Research Council, Washington DC, January 1967.

Research Studies Constituting the US Contribution to the International Biological Programme, Report no. 3 (Part I), National Research Council, Washington DC, June 1967.

Research Studies Constituting the US Contribution to the International Biological Programme, Report no. 3 (Part II), National Academy of Sciences, Washington, DC, December 1968.

US/IBP Annual Report, 1968, January 1969 (roneo).

Research Programs Constituting the US Participation in the IBP, Report no. 4. National Academy of Sciences, Washington DC, 1971.

US Participation in the IBP, July 1970–December 1971.

Report no. 5. National Academy of Sciences, Washington DC, 1973.

Report no. 6. National Academy of Sciences, Washington DC, 1975.

National synthesis now planned include: HA – 6 volumes on Man in the Andes, Eskimo of NW Alaska, Aleut adaptations, the Yanomama, Migration and Urbanization and Nutritional adaptation (in press).

Environmental Management – 25 volumes on biome and interbiome synthesis including 8 volumes on N. American grasslands (from 1974–6), 3 volumes on arid lands (1974–6), 2 volumes on Tundra (1975) and 2 volumes on woodlands (1974–5).

USSR

USSR Participation in the International Biological Programme, 1968. Academy of Sciences of the USSR, Leningrad.

Progress Report 1968 of the Soviet National Committee of IBP (USSR). Academy of Sciences of the USSR, Leningrad, 1969.

Progress Report 1971. Academy of Sciences of the USSR, Leningrad, 1973.

Estonian National Programme for IBP, Academy of Science of the Estonian SSR, Tartu 1970.

Estonian contribution to the IBP, Academy of Science of the Estonian SSR, Tartu 1970.

Estonian contribution to the IBP, II & III, Academy of Science of the Estonian SSR, Tartu, 1971.

Publications of IBP results by the Soviet National Committee now total 44 volumes (mainly in Russian) including: 18 volumes from Section PT including 4 volumes on tundra, 4 volumes on woodlands, 2 volumes on aridlands and 4 volumes on soil and soil processes; 4 volumes from PP on photosynthesis and nitrogen fixation; 13 volumes from PF; 3 volumes from PM; 4 volumes from HA; and 1 volume from UM. Further volumes, including one on grasslands, are in preparation.

The Academy of Sciences of the Estonian SSR have also published 5 volumes of results from Sections PP, PT, CT, PF & HA.

VENEZUELA

Venezuelan National Programme, Caracas, 1968.

YUGOSLAVIA

Preliminary Information on the National Programme of the Yugoslav National Committee for the IBP, April 1966 (roneo).

Yugoslav National Programme 1967.

5d. The international synthesis series

The evolution of IBP
Ed. E. B. Worthington (UK)

Woodlands: their structure and function (3 vols.)
Ed. D. E. Reichle (USA)

Grasslands: their structure and function
Ed. R. T. Coupland (Canada)

Appendix 5

Grasslands: comparative analysis
  Eds. A. Breymeyer (Poland) & G. M. Van Dyne (USA)

Tundra and related habitats
  Ed. J. J. Moore (Eire)

Aridlands: structure and utilization (2 vols.)
  Eds. R. A. Perry (Australia) & D. W. Goodall (Australia)

Small mammals: their productivity and population dynamics
  Eds. F. B. Golley (USA), K. Petrusewicz (Poland) & L. Ryszkowski
  (Poland)

Granivorous birds: their role and control
  Eds. S. C. Kendeigh (USA) & J. Pinowski (Poland)

Production ecology of ants and termites
  Ed. M. V. Brian (UK)

Ecology of wetlands
  Eds. J. Květ (Czechoslovakia), A. Szczepański (Poland) & D. G.
  Westlake (UK)

The functioning of freshwater ecosystems
  Ed. E. D. Le Cren (UK)

Photosynthesis and productivity in different environments
  Ed. J. P. Cooper (UK)

Nitrogen fixation by free-living micro-organisms
  Ed. W. D. P. Stewart (UK)

Symbiotic nitrogen fixation in plants
  Ed. P. S. Nutman (UK)

Decomposition and soil processes
  Ed. D. Parkinson (Canada)

Conservation of ecosystems
  Ed. A. R. Clapham (UK)

Crop genetic resources for today and tomorrow
  Eds. O. H. Frankel (Australia) & J. G. Hawkes (UK)

Studies in biological control
  Ed. V. Delucchi (Switzerland)

Food protein sources
  Ed. N. W. Pirie (UK)

Marine production mechanisms in different climatic zones
  Ed. M. J. Dunbar (Canada)

Man's influence on the marine environment
 Ed. J. B. Pearce (USA)

Marine mussels: their ecology and physiology
 Ed. B. L. Bayne (UK)

Aquaculture of mullets
 Ed. O. H. Oren (Israel)

Biology of circumpolar people
 Ed. F. A. Milan (USA)

Biology of high-altitude peoples
 Ed. P. T. Baker (USA)

World-wide variation in human growth
 Eds. P. B. Eveleth (UK) & J. M. Tanner (UK)

Components of human physiological function
 Ed. J. S. Weiner (UK)

Components of human population structure
 Ed. G. A. Harrison (UK)

QUANTITIES, UNITS AND SYMBOLS FOR IBP SYNTHESIS

Report of a SCIBP Working Group

The following international list of terms, symbols and units for use in environmental biology was first circulated as a typescript draft in July 1973 and after revision based on comments received, was issued as a preprint in August 1974, with the financial support of UNESCO. It met with an immediate demand and as a result of wide distribution, a number of minor errors and improvements have come to light, which are now incorporated. This appendix, not the preprint, is thus the authoritative version as at mid-1975. However, if it meets with acceptance by environmental biology, further additions or modifications will surely be required, so that the issue of revised versions of the list may be desirable in future.

International Council of Scientific Unions

SPECIAL COMMITTEE FOR THE INTERNATIONAL BIOLOGICAL PROGRAMME (SCIBP)

IBP Publications Committee, c/o Linnean Society
Burlington House, Piccadilly, London, WIV 0LQ, England

Published with assistance from UNESCO

CONTENTS

PREFACE

The SCIBP Bureau Meeting in London in March 1972 appointed a small working group to produce a list of terms, symbols and units which could be used consistently throughout the IBP synthesis volumes. In practice, this proved a more difficult task and took much longer than at first envisaged.

After considerable deliberation the working party decided that their recommendations must be based firmly on the International System of Units (SI). The working group believes that where this involves the abandonment of current practices the change should be made now so that the synthesis volumes published over several years will follow conventions likely to be current at the time that they will be read. This means that some quantities, units, and symbols regularly used throughout Phases I and II of IBP should now be abandoned.

It is also the case that some concepts used and quantities measured during the early stages of IBP have, often as a result of IBP research itself, been found wanting and thus must now be jettisoned unless they are to become encumbrances. A similar problem arises with those terms that usefully express temporary working ideas in informal discussion, but that cannot be defined rigorously enough for more formal communication: we have avoided listing such terms.

We have tried to include most of those quantities and symbols we believe will have widespread use in the IBP synthesis volumes, but have left out those which are more specialized. We hope that each volume will have a list of those terms that are peculiar to itself. For a few fields we have been unsuccessful in obtaining a list, either because of a shortage of time or willing experts, or because the field has not yet advanced to the stage where there is enough agreement over suitable terms and symbols.

The working group is very grateful to all those IBP workers from different sections and other experts, especially Mr W. G. Evans and Mr D. Tillotson of the Royal Society of London, who have given advice or helped in various ways. The other members of the group would also like to record their thanks to Dr K. L. Blaxter and Dr P. G. Jarvis who were largely responsible for the difficult and tedious task of collating and editing the provisional typescript, as well as that of assessing the various comments received following its circulation.

Finally SCIBP would like to express its gratitude to UNESCO for provision of financial support for this publication in preprinted form which it is hoped will gain wide circulation and prove a lasting contribution from IBP to developing programmes in biological and ecological sciences.

E. D. Le Cren (Chairman), K. L. Blaxter, Gina Douglas, P. G. Jarvis, A. Macfadyen, W. E. Ricker, A. Winter.

1. INTRODUCTION

This booklet has been designed to provide a common basis for the presentation of quantitative information arising from the International Biological Programme. It adopts, with minor modification, the now widely accepted International System of Units (SI).

The units that might be employed to express the various quantities in SI have been given so that the dimensions of the quantities are quite explicit.

The symbols for quantities are presented in a series of sections and some symbols appear in more than one section. This is unavoidable but it is thought that only very rarely will confusion arise.

The lists of symbols while they are extensive are not exhaustive.

Workers may wish to expand the lists in some instances when they should follow the general rules which are given.

Appendix 6

2. THE SI SYSTEM OF UNITS

The International System of Units (SI) comprises the SI *units* and the SI *prefixes*.

The SI units are of three kinds: *base, supplementary,* and *derived.* There are seven base units (see opposite page), one for each of the seven physical quantities: length, mass, time, electric current, thermo-dynamic temperature, luminous intensity, and amount of substance, which are regarded as dimensionally independent. There are two supple-mentary units (see page 210): one for plane angle and one for solid angle. The derived unit for any other physical quantity is that obtained by the dimensionally appropriate multiplication and division of the base units (see page 210). Fifteen of the derived units have special names and symbols.

Printing of symbols for units

The symbol for a unit should be printed in roman (upright) type, should remain unaltered in the plural, and should not be followed by a full stop except when it occurs at the end of a sentence.

Example: 5 cm but not 5 cms. and not 5 cm. and not 5 cms

The symbol for a unit derived from a proper name should begin with a capital roman (upright) letter.

Examples: J for joule and Hz for hertz

Any other symbol for a unit should be printed in lower case roman (upright) type.

Symbols for prefixes for units should be printed in roman (upright) type with no space between the prefix and the unit.

Multiplication and division of units

A product of two units may be represented in either of the ways:

$$\text{N m} \quad \text{or} \quad \text{N} \cdot \text{m}$$

The representation Nm is not recommended.

A quotient of two units may be represented in any of the ways:

$$\text{m s}^{-1} \quad \text{or} \quad \text{m/s} \quad \text{or} \quad \frac{\text{m}}{\text{s}} \quad \text{or} \quad \text{m} \cdot \text{s}^{-1}$$

but not ms^{-1}.

These rules may be extended to more complex groupings but more

208

than one solidus (/) should not be used in the same expression unless parentheses are used to eliminate ambiguity.

Examples: J K^{-1} mol^{-1} or J/(K mol) but not J/K/mol
cm^2 V^{-1} s^{-1} or (cm/s)/(V/cm) but not cm/s/V/cm

Definitions of the SI base units

metre The metre is the length equal to 1 650 763.73 wavelengths in vacuum of the radiation corresponding to the transition between the levels 2p$_{10}$ and 5d$_5$ of the krypton-86 atom.

kilogram The kilogram is the unit of mass; it is equal to the mass of the international prototype of the kilogram.

second The second is the duration of 9 192 631 770 periods of the radiation corresponding to the transition between the two hyperfine levels of the ground state of the caesium-133 atom.

ampere The ampere is that constant current which, if maintained in two straight parallel conductors of infinite length, of negligible circular cross-section, and placed 1 metre apart in vacuum, would produce between these conductors a force equal to 2×10^{-7} newton per metre of length.

kelvin The kelvin, unit of thermodynamic temperature, is the fraction 1/273.16 of the thermodynamic temperature of the triple point of water.

candela The candela is the luminous intensity, in the perpendicular direction, of a surface of 1/600 000 square metre of a black body at the temperature of freezing platinum under a pressure of 101 325 newtons per square metre.

mole The mole is the amount of substance of a system which contains as many elementary entities as there are atoms in 0.012 kilogram of carbon 12.

Note. When the mole is used, the elementary entities must be specified and may be atoms, molecules, ions, electrons, other particles, or specified groups of such particles.

Examples:
1 mole of HgCl has a mass equal to 0.236 04 kilogram
1 mole of Hg$_2$Cl$_2$ has a mass equal to 0.472 08 kilogram
1 mole of e$^-$ has a mass equal to 5.4860×10^{-7} kilogram
1 mole of a mixture containing $\frac{2}{3}$ mole of H$_2$ and $\frac{1}{3}$ mole of O$_2$ has a mass equal to 0.012 010 2 kilogram

209

Definitions of the SI supplementary units

radian The angle subtended at the centre of a circle by an arc of the circle equal in length to the radius of the circle.

steradian The solid angle subtended at the centre of a sphere by an area on the surface of the sphere equal in magnitude to the area of a square having sides equal in length to the radius of the sphere.

Names and symbols for SI base and supplementary units

Physical quantity	Name of SI unit	Symbol for SI unit
length	metre	m
mass	kilogram	kg
time	second	s
electric current	ampere	A
thermodynamic temperature	kelvin	K
luminous intensity	candela	cd
amount of substance	mole	mol
plane angle	radian	rad
solid angle	steradian	sr

Special names and symbols for SI derived units

Physical quantity	Name of SI unit	Symbol or SI unit	Definition of SI unit	Equivalent form(s) of SI unit
energy	joule	J	$m^2\,kg\,s^{-2}$	N m
force	newton	N	$m\,kg\,s^{-2}$	$J\,m^{-1}$
pressure	pascal	Pa	$m^{-1}\,kg\,s^{-2}$	$N\,m^{-2}$, $J\,m^{-3}$
power	watt	W	$m^2\,kg\,s^{-3}$	$J\,s^{-1}$
electric charge	coulomb	C	$s\,A$	A s
electric potential difference	volt	V	$m^2\,kg\,s^{-3}\,A^{-1}$	$J\,A^{-1}\,s^{-1}$, $J\,C^{-1}$
electric resistance	ohm	Ω	$m^2\,kg\,s^{-3}\,A^{-2}$	$V\,A^{-1}$
electric conductance	siemens	S	$m^{-2}\,kg^{-1}\,s^3\,A^2$	Ω^{-1}, $A\,V^{-1}$
electric capacitance	farad	F	$m^{-2}\,kg^{-1}\,s^4\,A^2$	$A\,s\,V^{-1}$, $C\,V^{-1}$
magnetic flux	weber	Wb	$m^2\,kg\,s^{-2}\,A^{-1}$	V s
inductance	henry	H	$m^2\,kg\,s^{-2}\,A^{-2}$	$V\,A^{-1}\,s$
magnetic flux density	tesla	T	$kg\,s^{-2}\,A^{-1}$	$V\,s\,m^{-2}$, $Wb\,m^{-2}$
luminous flux	lumen*	lm	cd sr	
illumination	lux*	lx	$m^{-2}\,cd\,sr$	
frequency	hertz	Hz	s^{-1}	

* In the definition here for these units, the steradian (sr) is treated as a base unit.

Examples of SI derived units and unit symbols for other quantities. (This list is merely illustrative)

Physical quantity	SI unit	A symbol for SI unit
area	square metre	m^2
volume	cubic metre	m^3
wavenumber	1 per metre	m^{-1}
radioactivity	1 per second	s^{-1}
density	kilogram per cubic metre	$kg\ m^{-3}$
speed; velocity	metre per second	$m\ s^{-1}$
acceleration	metre per second squared	$m\ s^{-2}$
kinematic viscosity	square metre per second	$m^2\ s^{-1}$
concentration (of amount of substance)	mole per cubic metre	$mol\ m^{-3}$
specific volume	cubic metre per kilogram	$m^3\ kg^{-1}$
molar volume	cubic metre per mole	$m^3\ mol^{-1}$
dynamic viscosity	pascal second	$Pa\ s$
surface tension	newton per metre	$N\ m^{-1}$
heat flux density	watt per square metre	$W\ m^{-2}$
heat capacity	joule per kelvin	$J\ K^{-1}$
thermal conductivity	watt per metre kelvin	$W\ m^{-1}\ K^{-1}$
energy density	joule per cubic metre	$J\ m^{-3}$
molar heat capacity	joule per kelvin mole	$J\ K^{-1}\ mol^{-1}$

SI prefixes

The following prefixes may be used to construct decimal multiples of units.

Multiple	Prefix	Symbol	Multiple	Prefix	Symbol
10^{-1}	deci	d	10	deca	da
10^{-2}	centi	c	10^2	hecto	h
10^{-3}	milli	m	10^3	kilo	k
10^{-6}	micro	μ	10^6	mega	M
10^{-9}	nano	n	10^9	giga	G
10^{-12}	pico	p	10^{12}	tera	T
10^{-15}	femto	f			
10^{-18}	atto	a			

Decimal multiples of the kilogram, kg, should be formed by attaching an SI prefix not to kg but to g, in spite of the kilogram and not the gram being the SI base unit.

Examples: mg not μkg for 10^{-6} kg
  Mg not kkg for 10^3 kg

Appendix 6

A symbol for an SI prefix may be attached to the symbol for an SI base unit, or for an SI supplementary unit, or for an SI derived unit having a special name and symbol.

Examples: cm ns μA mK μmol μrad

Compound prefixes should not be used.

Example: nm but not mμm for 10^{-9} m

A combination of prefix and symbol for a unit is regarded as a single symbol which may be raised to a power without the use of brackets.

Example: cm^2 always means $(0.01 \text{ m})^2$ and never 0.01 m^2

Decimal multiples of SI units having special names

(i)

Physical quantity	Name of unit	Symbol for unit	Definition of unit
area	hectare	ha	10^4 m^2
volume	litre	l	$10^{-3} \text{ m}^3 = dm^3$
mass	tonne	t	$10^3 \text{ kg} = Mg$

(ii) The following names are not part of SI. It is recognized that their use may be continued for some time but it is recommended that except in special circumstances they should be progressively abandoned in scientific publications. The following list is not exhaustive.

Physical quantity	Name of unit	Symbol for unit	Definition of unit
length	ångström	Å	$10^{-10} \text{ m} = 10^{-1} \text{ nm}$
length	micron	μm*	10^{-6} m
force	dyne	dyn	10^{-5} N
pressure	bar	bar	10^5 Pa
energy	erg	erg	10^{-7} J
kinematic viscosity	stokes	St	$10^{-4} \text{ m}^2 \text{ s}^{-1}$
dynamic viscosity	poise	P	10^{-1} Pa s

* The symbols μ and mμ, still unfortunately used by some spectroscopists and biologists, should give place to μm (micrometre) and nm (nanometre) respectively.

Other units now exactly defined in terms of the SI units

These units are not part of the SI. It is recognized that their use may be continued for some time, but it is recommended that except in special circumstances they should be progressively abandoned in scientific publications. Most of these units should not be used to form compound units. The following list is by no means exhaustive. Each of the definitions given in the fourth column is *exact*.

212

Physical quantity	Name of unit	Symbol for unit	Definition of unit
length	inch	in	2.54×10^{-2} m
mass	pound (avoirdupois)	lb	0.453 592 37 kg
time*	minute	min	60 s
	hour	h	60 min = 3 600 s
	day	d	24 h = 86 400 s
force	kilogram-force	kgf	9.806 65 N
pressure	atmosphere	atm	101 325 Pa
pressure	torr	Torr	(101 325/760) Pa [$\approx$ 133.322 368 Pa]
pressure	conventional millimetre of mercury	mmHg	$13.595\,1 \times 980.665 \times 10^{-2}$ Pa [$\approx$ 133.322 387 Pa]
energy	thermochemical calorie	cal_{th}	4.184 J
energy	I.T. calorie	cal_{IT}	4.1868 J
thermodynamic temperature	degree Rankine†	°R	(5/9) K
Celsius temperature *‡	degree Celsius†	°C	K
Fahrenheit temperature §	degree Fahrenheit†	°F	(5/9) K
radioactivity	curie	Ci	3.7×10^{10} s^{-1}
radiation**	rad††	rad	10^{-2} J kg^{-1}
	röntgen	R	2.58×10^{-4} C kg^{-1}

* Units for time and degree Celsius are now recognized by C.I.P.M. for continued use The recommendation that these should be progressively abandoned no longer applies See page 217 for further comments on time.

† The ° sign and the letter following form one symbol and there should be no space between them. *Example*: 25 °C not 25° C.

‡ The Celsius temperature is the excess of the thermodynamic temperature over 273.15 K.

§ The Fahrenheit temperature is the excess of the thermodynamic temperature over 459.67 °R.

** A special unit which takes account of the relative potentials for damage by different radiations and other factors is the rem (acronym for radiation equivalent man).

†† Whenever confusion with the symbol for the radian (angular measure) appears possible the symbol rd may be used for the rad.

Appendix 6

3. PRESENTATION OF NUMERICAL INFORMATION

Printing of numbers

Numbers should be printed in upright type. The decimal sign between digits in a number should be a point (.) or a comma (,). To facilitate the reading of long numbers the digits may be grouped in threes about the decimal sign but no point or comma should ever be used except for the decimal sign.

Example: 2573.421736 but not 2,573.421,736

When the decimal sign is placed before the first digit of a number a zero should always be placed before the decimal sign.

Example: 0.2573×10^4 but not $.2573 \times 10^4$

It is often convenient to print numbers with just one digit before the decimal sign.

Example: 2.573×10^3

Multiplication and division of numbers

The multiplication sign between numbers should be a cross ($\times$).

Example: 2.3×3.4

Division of one number by another may be indicated in any of the ways:

$$\frac{136}{273} \quad \text{or} \quad 136/273 \quad \text{or} \quad 136 \times (273)^{-1}$$

These rules may be extended to more complex groupings, but more than one solidus (/) should never be used in the same expression unless parentheses are used to eliminate ambiguity.

Examples: $(136/273)/2.303$ or $136/(273 \times 2.303)$
but never $136/273/2.303$

Numbers should be separated from other mathematical operators such as $+$, $-$ and $=$ by a space on each side of the operator (see section 10, page 248).

Numbers in figures and tables

Confusion sometimes arises in the interpretation of tables or figures where the numerical value of a quantity is a multiple of the unit used. For example, if the number in a table or on the axis of a graph is 1.0 and the definition is given as mass $\times 10^3$ in g, it is not immediately clear to all that the corresponding mass is 0.001 g and not 1000 g.

To obviate such difficulties, it is recommended that the numerical values of a quantity always be stated in the body of the table or on the axis of the graph explicitly. The clearest way to do this is to use the appropriate prefix, e.g. 1.0 mg, and this is recommended. Also acceptable but less succinct would be 1.0×10^{-3} g. Alternatively the column of the table or the axis of the graph should be labelled $10^3 \times$ mass/g or mass/mg.

Appendix 6

4. NOTES ON PARTICULAR ASPECTS OF SI

The calorie and the joule

The use of the joule as the unit of energy raises a series of problems about its relation to the calorie, for the calorie has been defined in different ways. The 15 °C calorie is defined as the amount of heat required to increase the temperature of water from 14.5 °C to 15.5 °C, the specific heat of water at 15 °C at constant pressure being defined as unity, while the Bunsen calorie is defined as 0.01 × the heat required to increase the temperature of water from 0 °C to 100 °C. Uncertainties about the joule equivalent of the Bunsen and the 15 °C calorie arise from uncertainties about the heat capacity of water over the temperature range considered. The thermochemical calorie is defined simply as 4.184 J, which in effect states that it is measured at that temperature at which the heat capacity at constant pressure of water is 4.184. The international steam table calorie is defined in electrical units giving a calorie equivalent of 4.1868 J.

There is every reason to identify the calorie used in studies in bio-energetics as the thermochemical calorie. The calibration of bomb calorimeters is undertaken using benzoic acid of thermochemical standard grade as a standard, this standard being expressed primarily in kJ/mol and converted to kcal/mol by dividing by 4.184.

The calorie has long been the unit of energy used in bioenergetic studies and many are naturally loth to change. The reasons for the desirability for change are, however, overriding. Firstly, adoption of the joule as the unit of energy provides a coherence for energy measurements in all those many branches of science on which biological studies ultimately depend. Secondly, to retain the calorie would lead to isolation of biological studies of energy transfer from the main body of advance in physics and chemistry.

The pascal

The pascal becomes the base unit of pressure. The bar may still be used but should be progressively abandoned (p. 10). However, the mbar and μbar should be dropped immediately. The conversions are as follows:

$$1 \text{ bar } = 10^5 \text{ Pa } = 0.1 \text{ MPa}$$
$$1 \text{ mbar } = 100 \text{ Pa } = 0.1 \text{ kPa}$$
$$1 \text{ μbar } = 0.1 \text{ Pa}$$

Time

For practical purposes, the SI unit of time, the second (s), is often inconveniently small. The following multiples of the second with special names may be used (see footnote * on page 213).

Name of unit	Symbol for unit	Definition of unit
minute	min	60 s
hour	h	3 600 s
day	d	86 400 s
week	wk	60.48×10^4 s
year	a	≈ 365 d

SI derived units (see page 210) that involve the dimension of time obviously use the base unit of time, the second. When dealing with power, however, it is sometimes inconvenient to use the derived SI unit, the watt. It is permissible to express power in terms of joules per unit time. For example, the rate of energy expenditure of an animal can be expressed in terms of $J\ d^{-1}$ rather than in terms of W.

In the same way, the components of the energy, mass (carbon, nutrients, etc.) and number balances of a system may be expressed on a per day or per year basis, particularly when the second is completely out of scale with the time period of the investigation and its use would imply a much higher accuracy than is realized.

Plane angle

In practice, the SI unit of plane angle, the radian (rad), is often replaced by the degree, minute and second. These units are not part of the SI but may be used in appropriate circumstances. Decimal fractions of the degree may be preferred to minutes and seconds in some circumstances.

Name of unit	Symbol for unit	Definition of unit
degree	°	$\pi/180$ rad
minute	′	$\pi/180 \times 60$ rad
second	″	$\pi/180 \times 3 600$ rad

Appendix 6

5. GENERAL RULES FOR QUANTITIES
Physical quantities

A physical quantity is defined by a complete specification of the operations used to measure the ratio (a pure number) of two particular values of that physical quantity.

Each physical quantity is given a name and a symbol which is an abbreviation for that name.

By international convention, seven physical quantities are chosen for use as dimensionally independent *base quantities*:

Physical quantity	Symbol for quantity
length	l
mass	m
time	t
electric current	I
thermodynamic temperature	T
luminous intensity	I_v
amount of substance	n

All other physical quantities are regarded as being *derived* from the base quantities.

Plane angle (symbol rad) and solid angle (symbol sr) are sometimes regarded as base quantities.

Symbols for physical quantities

The symbol for a physical quantity should be a single letter of the latin or the greek alphabet.

An exception to this rule has been made for certain dimensionless quantities used in the study of transport processes, for which the internationally agreed symbols consist of two letters, the first a capital and the second lower case. *Example*: Reynolds number *Re*. When necessary to avoid ambiguity, such two-letter symbols should be enclosed in parentheses.

When necessary the symbol for a physical quantity may be modified by attaching to it subscripts and/or superscripts and/or other modifying signs having specified meanings.

218

Printing of symbols for physical quantities

When letters of the latin alphabet are used as symbols for physical quantities they should be printed in italic type. When letters of the greek alphabet are used as symbols for physical quantities they should whenever possible be printed in sloping (italic) rather than upright (roman) type.

A product of two quantities may be represented in any of the ways:

$$ab \quad \text{or} \quad a\,b \quad \text{or} \quad a \cdot b \quad \text{or} \quad a \times b$$

A space should be left on each side of other mathematical operators such as $+$, $-$ and $=$ (see section 10).

Choice of symbols for physical quantities

Lists of recommended symbols for physical quantities follow. Whenever possible the symbol used for a physical quantity should be that (or one of those) recommended there.

Even with the use of both capital and lower case letters, the available distinctive letter symbols are insufficient to enable each symbol to be allotted to a single quantity. Some alternatives are therefore given in the lists where a need for them is most likely to arise or, occasionally, where alternative usages are firmly established and unobjectionable. In some instances a preference is expressed and the preferred symbol should then be used whenever possible.

Where it is necessary to choose from alternative symbols for a quantity, or to adopt a symbol for a quantity not listed, consideration should be given to current practice by authorities in the field and to the desirability that symbols for quantities constituting a well defined class should as far as possible belong to the same alphabet, fount, and case.

Modifying of signs

Letter symbols, numbers, or other signs, may be placed as subscripts or superscripts immediately before or after the symbol for a physical quantity in order to modify its meaning.

For the use of subscripts and superscripts, and of other modifying signs, no rigid rules are laid down but a satisfactory notation should fulfil the following requirements:

   (i) it should be unambiguous;

   (ii) it should be simple, systematic, and easy to remember;

   (iii) it should not use more letters than necessary;

   (iv) it should not be too expensive or difficult to print.

Appendix 6

Modifying signs such as dots, bars, or tildes ($\sim$) may be placed above (or exceptionally below) the symbol for a physical quantity. Such signs, however, should be used sparingly and should never be letters of the alphabet or numbers.

Brackets, including parentheses ( ), braces { }, square brackets [ ], and angle brackets $\langle \rangle$, should not be used around the symbol for a quantity in order to make it represent any other quantity, unless such use is consistently adopted for a whole class of quantities as in crystallography. In particular, the use of square brackets around a chemical formula to denote the concentration of the substance is recommended.

Printing of subscripts and superscripts

Subscripts or superscripts which are themselves symbols for physical quantities should be printed in italic (sloping) type. All other letter symbols, and all numbers, used as subscripts or superscripts should be printed in roman (upright) type.

Example: C_p for heat capacity at constant pressure, but C_B for heat capacity of substance B

When two or more subscripts, or two or more superscripts, having separate meanings are attached to the same symbol they should be separated by commas.

Example: $C_{p,B}$ for heat capacity at constant pressure of substance B

Second-order superscripts or subscripts should be avoided as far as possible. Thus e^{x^2} may be printed as $\exp x^2$. Also $A_{NO_3^-}$ may be printed as $A(NO_3^-)$ and $\rho_{20\,°C}$ as $\rho(20\ °C)$.

Use of the words 'specific' and 'molar'

The word 'specific' before the name of an extensive physical quantity is restricted to the meaning 'divided by mass'. For example, specific volume is the volume divided by the mass. When the extensive quantity is represented by a capital letter, the corresponding specific quantity may be represented by the corresponding lower case letter.

Examples: volume: V specific volume: $v = V/m$
heat capacity: C_p specific heat capacity: $c_p = C_p/m$

The numerical value of a specific physical quantity depends on the units selected for the physical quantity and for the mass.

The word 'molar' before the name of an extensive quantity is restricted to the meaning 'divided by amount of substance'. For example, molar volume is the volume divided by the amount of substance. The subscript

220

m attached to the symbol for the extensive quantity denotes the corresponding molar quantity.

Examples: volume: V molar volume: $V_m = V/n$

Gibbs function: G molar Gibbs function: $G_m = G/n$.

The subscript m may be omitted where there is no risk of ambiguity.

The numerical value of a molar physical quantity depends on the units selected for the physical quantity and for the amount of substance. The most commonly used unit for amount of substance is the mole (see page 210).

Appendix 6

6. RECOMMENDED SYMBOLS FOR PHYSICAL QUANTITIES

It is recognized that in the lists which follow some symbols are used to describe more than one quantity. This is unavoidable but it is thought that in most instances no ambiguity will arise.

6a. *Space, time and mass*

Quantity	Symbol	Symbol for unit
angle (plane angle)	$\alpha, \beta, \gamma, \theta, \phi$, etc.	rad
solid angle	Ω, ω	sr
length	l	m
breadth	b	m
height, depth (dimension in the vertical)	h, z	m
thickness	d, δ	m
radius	r	m
diameter: $2r$	d	m
distance along path	s, L	m
generalized coordinate	q	m
rectangular coordinates	(x, y, z)	m
area (see page 9)	A, S	m²
volume	V, v	m³
time	t	s
velocity: ds/dt	(u, v, w)	m s⁻¹
acceleration: du/dt	a	m s⁻²
mass	m	kg
density, mass density: m/V	ρ	kg m⁻³
specific volume: V/m	v	m³ kg⁻¹
force	F	N
pressure	p, P	Pa, N m⁻²

6b. Light and radiation

The same symbol is used for the corresponding luminous and energetic quantity; the subscript e for energetic and v for visible may be used where confusion might otherwise occur.

Quantity	Symbol	Symbol for unit
radiant energy	Q, Q_e	J
radiant flux: dQ_e/dt (rate of propagation of radiant energy)	Φ, Φ_e	W, J s^{-1}
radiant flux density: $d\Phi_e/(dS \cos \alpha)$ (radiant flux passing through a plane of unit area)	F, F_e	W m^{-2}
radiant intensity: $d\Phi_e/d\omega$ (radiant power emitted by a source per unit solid angle)	I, I_e	W sr^{-1}
radiance: $d\Phi_e/(d\omega\,dS \cos \alpha)$ (radiant flux intensity per unit area in direction of emission)	L, L_e	W m^{-2} sr^{-1}
irradiance: $d\Phi_e/dS$* (incident radiant flux per unit area)	$_eI, _eI_e$	W m^{-2}, J m^{-2} s^{-1}
radiant exposure: dQ_e/dS	H, H_e	J m^{-2}
luminous intensity: $d\Phi_v/d\omega$	I, I_v	cd
luminous flux: dQ_v/dt	Φ, Φ_v	lm, cd sr
luminous flux density: $d\Phi_v/(dS \cos \alpha)$	F, F_v	lm m^{-2}, lx
quantity of light: $\int\Phi_v\,dt$	Q, Q_v	lm s
luminance, brightness: $d\Phi_v/(d\omega\,dS \cos \alpha)$	L, L_v	lm sr^{-1} m^{-2}, cd m^{-2}
illuminance: $d\Phi/dS$*	$_eI, _eI_v$	lm m^{-2}, lx
light exposure: $\int E_v\,dt$	H, H_v	lx s
flux of long-wave radiation per unit area (subscript u upward; d downward; e from environment; b from body)	L	W m^{-2}
net radiation flux density	R_n	W m^{-2}

* Where confusion with radiant or luminous intensity will not occur the prefix e may be dropped.

Appendix 6

Quantity	*Symbol*	*Symbol for unit*
solar irradiance on a horizontal surface (subscript b for direct beam; d for diffuse)	$_eI$	W m^{-2}
direct solar irradiance on surface perpendicular (normal) to solar beam	$_eI_p$	W m^{-2}
solar radiation received by a body per unit area as a result of reflection from the environment	$_eI_r$	W m^{-2}
solar zenith angle	α	
solar elevation	β	
irradiance at wavelength λ	$_eI(\lambda)$	W m^{-2}
photosynthetically active irradiance (range of λ in nm, e.g. 380–720)	$_eI(\lambda_1 - \lambda_2)$	W m^{-2}
photon flux (incident flux of photons per unit area)	$_eI_q$	*Einstein m^{-2} s^{-1}
fraction of sky covered by cloud	c	dimensionless
absorptance, absorption factor (ratio of absorbed to incident radiant flux)	α	dimensionless
absorptivity at wavelength λ	$\alpha(\lambda)$	dimensionless
emittance (ratio of emitted radiant flux to that from a black body at same temperature)	ϵ	dimensionless
emissivity at wavelength λ (subscript a apparent emissivity of the atmosphere)	$\epsilon(\lambda)$	dimensionless
reflectance, reflection factor (ratio of reflected to incident radiant flux)	ρ	dimensionless
reflectivity at wavelength λ	$\rho(\lambda)$	dimensionless
transmittance, transmission factor (ratio of transmitted to incident radiant flux)	τ	dimensionless

* The Einstein is a mole of photons; i.e. Avogadro's constant (6.022169×10^{23}) of photons.

224

Quantity	Symbol	Symbol for unit
internal transmittance (transmittance of the medium itself disregarding boundary or container influence)	τ_i	dimensionless
absorbance: $\log_{10} (1/\tau_i)$ (subscript λ wavelength, x pigment)	$D_{i,\,x}A_\lambda$	dimensionless
absorption coefficient (D_i/l)	a	m^{-1}
molar absorption coefficient (a/molar concentration)	a_m	$m^2\,mol^{-1}$
specific absorption coefficient (a/mass concentration)	$a_s,\,\alpha$	$m^2\,kg^{-1}$
extinction or attenuation coefficients		
in foliage: $I = I_0 \exp(-KL)$	K	dimensionless
in water: $I = I_0 \exp(-\epsilon z)$	ϵ	m^{-1}
frequency	$f,\,\nu$	s^{-1}
wavelength	λ	m, nm
wave number	ν	m^{-1}
speed of light in vacuum	c	$m\,s^{-1}$

Spectra

For all kinds of photobiological spectra the abscissa should be a wavelength scale with low values to the left. The ordinates should be as follows:

action spectra: action per photon (or einstein) but not per unit energy.

emission spectra: photons (or einstein) per unit wavelength interval, or, if there are special reasons, energy per unit wavelength interval.

absorption spectra: absorbance, or, if there are special reasons, transmittance or absorptance.

6c. Heat exchange

Quantity	Symbol	Symbol for unit
temperature (subscripts a ambient, b body, l leaf, etc.)	T	K, °C
heat, quantity of heat	Q	J
work, quantity of work	W	J

Appendix 6

Quantity	Symbol	Symbol for unit
heat flux: dQ/dt (subscripts r radiation, c convection, k conduction, e evaporation, s in or out of storage)*	H	W, J s^{-1}
heat flux per unit area: $dQ/(S\,dt)$ (subscripts as above using rule on page 220)	H_S	W m^{-2}
rate of metabolic heat production $(= H_r + H_c + H_k + H_e + H_s)$	H_M	W, J s^{-1}
specific energy, enthalpy of combustion per unit mass: Q/m (subscripts t tissue, w wood, etc.)	J	J kg^{-1}
specific heat capacity at constant volume	c_v	J kg^{-1} K^{-1}
specific heat capacity at constant pressure	c_p	J kg^{-1} K^{-1}
specific heat capacities of biological entities (subscripts s soil, t tissue, w wood)	c	J kg^{-1} K^{-1}
thermal conductance or heat transfer coefficient: $dQ/(S\,dt)$ per unit temperature difference (for surface heat transfer the subscripts r radiation, c convection, k conduction)	h	W m^{-2} K^{-1}
thermal conductivity, rate of conductive transfer of heat across unit area for unit temperature gradient: $dQ/(S\,dT(dt/dx))$	k	W m^{-1} K^{-1}
insulation, reciprocal of thermal conductance: $1/h$ (subscripts t tissue, a air in surface boundary layer, f hair coat)	I	K W^{-1} m^2
resistance to heat transfer (subscripts as above)	r_h	s m^{-1}

---

* In some instances it may prove convenient to state the components of the heat balance simply as r, c, k, etc. rather than H_c, H_r, H_k, etc. The fact that this simplification is adopted should, however, be stated.

226

Quantity	Symbol	Symbol for unit
thermal diffusivity (subscripts as above)	κ	$\mathrm{m^2\,s^{-1}}$
transfer coefficient for turbulent heat transfer in air	K_h	$\mathrm{m^2\,s^{-1}}$
Grashof number	Gr	dimensionless
Nusselt number	Nu	dimensionless
Prandtl number: ν/κ	Pr	dimensionless

6d. Water vapour exchange

(General subscripts v, H_2O to indicate water vapour, may be omitted where there is no risk of ambiguity.)

Quantity	Symbol	Symbol for unit
partial pressure of water vapour in air (subscripts a in ambient air, i at liquid–air interface, etc.)	e, P_{H_2O}	Pa
saturation water vapour pressure at temperature T	$e_s\,(T)$	Pa
saturation deficit, vapour pressure deficit: $e_s\,(T)-e_a$	δe	Pa
leaf–air vapour pressure deficit	δe	Pa
relative humidity: $100e/e_s\,(T_a)$	r.h.*	dimensionless
absolute humidity of air (subscripts as above)	χ	$\mathrm{kg\,m^{-3}}$
saturation absolute humidity at temperature T	$\chi_s\,(T)$	$\mathrm{kg\,m^{-3}}$
specific humidity of air (subscripts as above)	q	$\mathrm{kg\,kg^{-1}}$
saturation specific humidity at temperature T	$q_s\,(T)$	$\mathrm{kg\,kg^{-1}}$
flux density of water vapour in evaporation (subscripts e from leaves, c from canopy, g from ground, s from skin, r in respiration by an animal, o from open water, i.e. potential)	E	$\mathrm{kg\,m^{-2}\,s^{-1}}$

* Abbreviation, not symbol.

Quantity	*Symbol*	*Symbol for unit*
latent heat of vaporization of water	L, λ	$J\ kg^{-1}$
flux density of latent heat: $L \cdot E, \lambda \cdot E$ (subscripts as above)	H_e	$W\ m^{-2}$
diffusion coefficient for water vapour in air (subscripts o free diffusion coefficient, e effective diffusion coefficient)	D_v, D_{H_2O}	$m^2\ s^{-1}$
transfer coefficient for turbulent transport of water vapour in air	K_v, K_{H_2O}	$m^2\ s^{-1}$
resistance to water vapour transfer (subscripts a in atmosphere across surface boundary layer, c of cuticle, i of intercellular spaces, p of pore, s of stomata, l of leaf, ab, ad of abaxial and adaxial surfaces, respectively, c of canopy, b of body, f of hair, clothing, etc.)	r_v, r_{H_2O}	$s\ m^{-1}$
conductance to water vapour transfer (subscripts as above)	k_v, k_{H_2O}	$m\ s^{-1}$
psychrometric constant	γ	$Pa\ K^{-1}$,
wet-bulb temperature	T_w	K, °C
equivalent temperature: $T + e/\gamma$	θ	K, °C
rate of change of saturation vapour pressure with temperature: $\partial e_s(T)/\partial T$	s, Δ	$Pa\ K^{-1}$,
rate of change of latent heat content with sensible heat content of saturated air: $\partial e_s(T)/(\partial T\ \gamma)$	ϵ	dimensionless
ratio of flux densities of convective and latent heat (Bowen ratio): H_c/H_e	β, B	dimensionless
Lewis number κ/D	Le	dimensionless
Schmidt number	Sc	dimensionless
Sherwood number	Sh	dimensionless

6e. *Carbon dioxide and oxygen exchange, and photosynthesis*

(General subscripts CO_2 and O_2, and A, S or V may be omitted where there is no risk of ambiguity; multiple subscripts follow the rules given in section 5.)

Quantity	Symbol	Symbol for unit
concentration of CO_2	c_{CO_2}	mol m^{-3}
concentration of O_2	c_{O_2}	mol m^{-3}
mass concentration of CO_2	ρ_{CO_2}	kg m^{-3}
mass concentration of O_2	ρ_{O_2}	kg m^{-3}
volume fraction of CO_2	ϕ_{CO_2}	dimensionless
volume fraction of O_2	ϕ_{O_2}	dimensionless
volume concentration of CO_2	C_{CO_2}	cm^3 m^{-3}
partial pressure of CO_2	P_{CO_2}	Pa
partial pressure of O_2	P_{O_2}	Pa
(subscripts for all above: a in ambient air, c in canopy, i mean intercellular space, x at carboxylation site, etc.)		
diffusion coefficient of CO_2 in air	$_aD_{CO_2}$	m^2 s^{-1}
diffusion coefficient of CO_2 in water	$_wD_{CO_2}$	m^2 s^{-1}
diffusion coefficient of O_2 in air	$_aD_{O_2}$	m^2 s^{-1}
diffusion coefficient of O_2 in water	$_wD_{O_2}$	m^2 s^{-1}
turbulent transfer coefficients		
for CO_2 in air	$_aK_{CO_2}$	m^2 s^{-1}
for CO_2 in water	$_wK_{CO_2}$	m^2 s^{-1}
for O_2 in air	$_aK_{O_2}$	m^2 s^{-1}
for O_2 in water	$_wK_{O_2}$	m^2 s^{-1}
resistance to CO_2 transfer	r_{CO_2}	s m^{-1}
(subscripts as for r_v, r_h; m in the mesophyll cells; x carboxylation; e excitation)		
conductance for CO_2 transfer	k_{CO_2}	m s^{-1}
(subscripts as above)		
nuclide of carbon of mass number 14	^{14}C	—
nuclide of oxygen of mass number 18	$^{18}O_2$	—
CO_2 compensation concentration	Γ	cm^3 m^{-3}

Appendix 6

Quantity	Symbol	Symbol for unit
average energy requirement for the photosynthetic fixation of CO_2	γ, ϵ	$J\,mol^{-1}$, $J\,kg^{-1}$
quantum yield (see page 244)	Φ	dimensionless (mol Einstein^{-1})
concentration of chlorophyll	c_a, c_b, $c_{(a+b)}$	$mol\,m^{-2}$, $mol\,m^{-3}$, $mol\,kg^{-1}$ ($kg\,m^{-3}$)

(subscripts A per unit ground/water area: m^{-2}, S per unit surface area of organ or organism: m^{-2}, V per unit volume of habitat or organism: m^{-3}, m per unit mass of organ or organism: kg^{-1}, e.g. $_S c_a$)

Quantity	Symbol	Symbol for unit
rate of net gas exchange measured as a flux of CO_2 flux of O_2	$\left.\begin{array}{c} F_{CO_2} \\ F_{O_2} \end{array}\right\}$	$mol\,s^{-1}$ ($m^3\,s^{-1}$, $kg\,s^{-1}$)

(subscript i influx, e efflux, l leaf, s stem, r root, p plant, a animal, A per unit ground/water area: m^{-2}, S per unit surface area of organism: m^{-2}, V per unit volume of habitat or organism: m^{-3}, m per unit mass of organism: kg^{-3}, result in a change in the units, e.g. $_A F_{CO_2}$ has units of $mol\,m^{-2}\,s^{-1}$)

Quantity	Symbol	Symbol for unit
flux of CO_2 or O_2 per unit amount of chlorophyll	$_c F$	s^{-1}
average rate of net photosynthesis or respiration (positive or negative) in the interval of time $t - t_0 = \Delta t$ determined as a change		
in individual mass	F_m	$kg\,s^{-1}$
in individual energy	F_Q	$J\,s^{-1}$, W
in biomass	F_B	$kg\,s^{-1}$
in energy of biomass	F_A	$J\,s^{-1}$, W
(subscripts and superscripts as above)		

6f. Momentum transfer

(General subscript M may be omitted where there is no risk of ambiguity.)

Quantity	Symbol	Symbol for unit
velocity of air at height z above Earth's surface in the x, y, z directions	$u(z)$, $v(z)$, $w(z)$	m s^{-1}
friction velocity	u_*	m s^{-1}
flux density of momentum, shearing stress	τ	N m^{-2}
transfer coefficient for turbulent transport of momentum in air	K_M	m^2 s^{-1}
resistance to momentum transfer (subscript a in the atmosphere across surface boundary layer, c of canopy boundary layer)	r_M	s m^{-1}
drag coefficient for form drag and skin friction combined	C_d	dimensionless
drag coefficient for form drag alone	C_f	dimensionless
zero plane displacement	d	m
height above Earth's surface	z	m
roughness length	z_0	m
von Kármán's constant	k	dimensionless
height of equilibrium boundary layer	Z	m
depth of a boundary layer	δ	m
coefficient of dynamic viscosity of air	μ	Pa s
coefficient of kinematic viscosity of air	ν	m^2 s^{-1}
ratio of projected area of leaves (S_l) to frontal (outline) area of shoot	σ	dimensionless
shelter factor, ratio of exchange coefficients of isolated leaves and leaves on the plant (subscripts d drag, v water vapour, etc.)	p	dimensionless
Reynolds number	Re	dimensionless
Richardson number	Ri	dimensionless

6g. Soil and plant water

(Subscripts s soil, l leaf, x xylem, r root, t at full turgor.)

Quantity	Symbol	Symbol for unit
volume fraction of water	ϕ	dimensionless
mass fraction of water	θ	dimensionless
mass concentration of water	ρ_w	kg m^{-3}
dynamic viscosity of water	μ	Pa s
partial molar volume of water	v_{H_2O}	m^3 mol^{-1}
volume of water in sample (subscript t at full turgor, e expressed, d deficit, o of free water, b of bound water)	V	m^3
relative water content: $(V_t - V_d)/V_t$	R	dimensionless
free water content: $(V_o - V_d)/V_o$	F	dimensionless
bound water content: $(V_t - V_o)/V_o$	B	dimensionless
total potential: $RT\ln(e/e_s)/v_{H_2O}$	ψ	J m^{-3}, Pa
matric potential	τ	J m^{-3}, Pa
solute potential	π	J m^{-3}, Pa
pressure (turgor) potential	P	J m^{-3}, Pa
gravitational potential	P_g	J m^{-3}, Pa
submergence potential	P_u	J m^{-3}, Pa
volume flux	q	m^3 s^{-1}
volume flux density: q/s	v	m^3 m^{-2} s^{-1}
hydraulic conductivity of soil: $v/(d\psi/dz)$	K	s
diffusivity of water in soil: $K(d\psi/d\phi)/\rho_w$	D_l	m^2 s^{-1}
hydraulic resistivity of plant: $(d\psi/dz)/(v\rho_w)$	R_p	s^{-1}
hydraulic resistance of plant: $\Delta\psi/(qp_w)$	R_p	m^{-1} s^{-1}
hydraulic conductivity of plant: $1/\rho$	σ	s
hydraulic conductance of plant: $1/R_p$	K_p	m s
relative conductivity, permeability constant: $v\mu/(d\psi/dz)$	k	m^2

7. SYMBOLS AND UNITS FOR CHEMICAL AND BIO-CHEMICAL QUANTITIES

Quantity	*Symbol*	*Symbol for unit*
Molecular quantities		
relative atomic mass of an element (atomic mass)	A_r	dimensionless
relative molecular mass of a substance (molecular mass)	M_r	dimensionless
molar mass (subscripts a dry air, v water vapour, H_2O, CO_2, etc.)	M	kg mol^{-1}
number of molecules	N	—
amount of substance	n	mol
mole fraction of substance B: $n_B/\Sigma_i n_i$	x_B	dimensionless
mass fraction of substance B	w_B	dimensionless
volume fraction of substance B	ϕ_B	dimensionless
molality of solute B: n_B/mass of solvent	m_B	mol kg^{-1}
concentration of solute B: n_B/V (formerly called molarity)	c_B	mol m^{-3}
mass concentration of substance B	ρ_B	kg m^{-3}
Rates of reaction		
stoichiometric coefficient of substance B	ν_B	dimensionless
extent of reaction: $d\xi = dn_B/\nu_B$	ξ	mol
rate of reaction: $d\xi/dt$	ξ, J	mol s^{-1}
rate of increase in concentration of substance B: dc_B/dt	v_B, r_B	mol m^{-3} s^{-1}
rate constant	k	m^3 s^{-1}
equilibrium constant	K	mol m^{-3}
Michaelis–Menten constant	K_m	mol m^{-3}

Appendix 6

8. SYMBOLS AND UNITS FOR QUANTITIES IN STUDIES OF BIOLOGICAL PRODUCTIVITY

The symbols given in this section are presented in an order. First are those quantities that are common to all studies. Included here is a distinction between A, the surface area of land and/or water, and S, the surface area of a plant, animal or a part of a plant or animal. Similarly the symbol V is reserved for the volume of a water, soil or air habitat and v for volume of plants, animals and their components. Second are those quantities that relate to certain concepts in the study of productivity in both plants and animals and in ecosystems. A division has been made between those quantities that are integral ones and those that are time rates.

The symbols for the unit are in the base units of SI. In many studies, however, the second as the unit of time will not be used and the day or the year will be employed.

Production and Productivity

The term production is used to conceptualize the total amount of biological material which is synthesized by a species or group of species.

Production is an amount of substance or energy. In most biological studies, however, rates of production are considered, and while in common speech the term production might be used to describe rate of production, in written work the term rate of production (kg s^{-1}, J s^{-1}) should be used to avoid ambiguity. Rates of production are usually expressed per unit of land or water area, per unit volume of water or per individual animal or plant, and these should be stated explicitly.

The term productivity is a non-quantitative term used to describe in a general way the summation of all those processes which are concerned in the production of biological material.

Some further expansion of the concept of production rate is necessary. The rate of change in the mass of a population of a species per unit time, that is the average growth rate of biomass, is not usually synonymous with production because some of the material synthesized during the time interval may have been removed or transferred to another species. Production rate is thus more rigorously defined as the rate of change of biomass plus the rate at which biomass, as individuals or parts of individuals, has been removed by death and subsequent microbial or other dissolution, by grazing, predation, cropping or indeed by any other physical or biological process.

234

Because of the multiplicity of ways in which material synthesized in an interval of time might be removed, it is important that when, in any publication, the term production rate is used it should be clearly stated what channels of removal of biomass have been considered. For example, the production rate of a meadow in a time interval could be expressed as the change in biomass in unit time plus the amount consumed per unit time by a species of large herbivore. Such an estimate ignores the rates of consumption by other species, by microbial decay and by shedding and death of leaves and roots.

A further example relates to the use of the term 'gross production', which, for the following reasons, is not a term to be retained.

Conceptually, gross photosynthesis is an unsatisfactory term because the respiration which occurs in the light during photosynthesis is very closely coupled to the photosynthetic process. For example, photorespiration has the same action spectrum as photosynthesis and its rate is linked to the rate of photosynthesis. Recent research has led to the conclusion that the carboxylating enzyme, RuDP carboxylase (EC 4.1.1.39), also acts as an oxygenase and produces the primary substrate for photorespiration (phosphoglycolate), concurrently with the first products of CO_2 fixation.

Practically, it is unsatisfactory because the plant respiration that occurs in the light concurrently with photosynthesis cannot be measured or reliably estimated on a routine basis. This respiration is therefore usually ignored in production studies. Consequently, for operational purposes, gross photosynthesis or gross production is usually taken as the absolute sum of net photosynthesis in the light and respiration in the dark of the plant or stand. This is not gross production in the sense implied by the use of the word 'gross' and in terrestrial C_3-plants is a considerable underestimation because of photorespiration.

Thus photorespiration is an inevitable consequence of photosynthesis from which it cannot be separated either practically, as an amount, or conceptually, as an independent process. Hence gross photosynthesis and gross production are not viable terms. The continued use of a quantity called gross production, based on an operational definition, might be justified on the grounds of familiarity of usage. However, such usage leads to apologies by the user and confusion of the recipient. Consequently the term should be dropped altogether and any similar, single quantity required should be defined in terms of the rate processes or amounts being summed (e.g. net photosynthesis of leaves + respiration of leaves at night + respiration of non-photosynthetic tissue (stems, roots, etc.) during day and night). A single symbol can be given to this sum if one is needed for purposes of discussion, but it should not be called gross photosynthesis or gross production.

This problem arises in particular in plankton biology where the net exchanges of O_2 or CO_2 are measured simultaneously in clear and darkened bottles suspended in the water. Usual practice is to add together the absolute fluxes in the two bottles to give an estimate of gross photosynthesis. This estimate is based on certain implicit, not necessarily correct, assumptions.

For example the observed net exchange of CO_2 (over say 3 h) in the light, $_lF_o$, is

$$_lF_o = {_lF_{i,p}} + {_lF_{e,p}} + {_lF_{e,a}} \tag{1}$$

(subscripts l light, o observed, i influx (+ve), e efflux (−ve), p plant, a animal).

The observed net exchange of CO_2 in the dark, $_dF_o$, is

$$_dF_o = {_dF_{e,p}} + {_dF_{e,a}}. \tag{2}$$

Subtracting equation (2) from (1) gives

$$_lF_o - {_dF_o} = {_lF_{i,p}} + {_lF_{e,p}} + {_lF_{e,a}} - {_dF_{e,p}} - {_dF_{e,a}}.$$

It is reasonable to assume that the animal respiration is the same in both light and dark, i.e. that $_lF_{e,a} = {_dF_{e,a}}$, so that

$$_lF_o - {_dF_o} = {_lF_{i,p}} + {_lF_{e,p}} - {_dF_{e,p}},$$

and if it is assumed that the rates of light and dark respiration of the algae are equal, then

$$_lF_o - {_dF_o} = {_lF_{i,p}}.$$

However, this last assumption may be far from correct ($_lF_{e,p} > {_dF_{e,p}}$) and hence an estimate of $_lF_{i,p}$ derived in this way is often a minimum estimate of $_lF_{i,p}$ which may be called the *photo-assimilation rate* and given the symbol A_p, if it is required.

As far as animals (and bacteria) are concerned, *assimilation* is also unsatisfactory because it is difficult both to define and measure precisely but it is nevertheless a useful ecological concept. It should be used only for *consumption* less *faeces*, i.e. the quantity of matter or energy which becomes incorporated into the animal (the gut being outside the animal). Problems arise in being precise about secretions into the gut, measuring faeces separately from excretions (urine) and the extent to which all excretion has at one time been incorporated into the animal. In all cases the exact observational or experimental procedures should be stated.

8a. Primary and secondary productivity

Quantity	Symbol	Symbol for unit
General		
area of land or water to which estimates of production are referred	A	m²
area of surface of an individual plant or part of plant or animal or part of animal (subscript 1 area of one leaf surface; 2l area of both (total) surfaces)	S	m²
volume of soil, water or air to which estimates of production are referred (subscripts g soil, w water, c canopy)	V	m³
volume of plant(s), animal(s) or their parts (subscripts s stem, a animal, etc.)	v	m³
time*	t	s
discrete interval of time $(t - t_0)$	Δt	s
designation of a particular species or group of species in an ecosystem (levels designated 1, 2, 3, ... j)	Subscript before the quantity symbol, i.e. $_j m_{Ca}$ (mass of calcium in an individual organism in species j at a particular time)	
mass of a complete individual organism or of a component part of an individual organism, at a particular time (subscripts (1) l leaf, r root, s shoot, g gonadic material, etc.; (2) D dry mass, C carbon, Ca calcium, c chlorophyll, etc.)	m	kg
energy (enthalpy of combustion) of an individual organism or of a component part of an individual organism at a particular time (subscripts l leaf, r root, s shoot, g gonadic material, etc.)	Q	J

* See page 217

Appendix 6

Quantity	Symbol	Symbol for unit
number of organisms in a population at a particular time	N	
biomass: mass of population of organisms (mN) or parts of organisms, at a particular time	B	kg
(subscripts (1) l leaf, r root, s shoot, g gonadic material, etc.; (2) D dry mass, C carbon, Ca calcium, c chlorophyll, etc.)		
energy (enthalpy of combustion) of biomass (QN)	Λ	J
(subscripts as in (1) above)		

mean value of x during time interval $t - t_0 = \Delta t$

$$\hat{x} = \frac{1}{\Delta t} \int x \, \mathrm{d}t$$

in individual mass	$\hat{m}$	kg
in individual energy	$\hat{Q}$	J
in biomass	$\hat{B}$	kg
in energy of biomass	$\hat{\Lambda}$	J
in numbers	$\hat{N}$	—

mean value for an individual organism in a population at a particular time

$$\bar{y} = \frac{1}{N} \sum_1^N y$$

mean individual mass	$\bar{m}$	kg
mean individual energy	$\bar{Q}$	J
population 'density'		
number per unit area: N/A	$_A N$	m^{-2}
number per unit volume: N/V	$_V N$	m^{-3}
biomass 'density'		
per unit area: B/A	$_A B$	kg m^{-2}
per unit volume: B/V	$_V B$	kg m^{-3}
energy of biomass		
per unit area: Λ/A	$_A \Lambda$	J m^{-2}
per unit volume: Λ/V	$_V \Lambda$	J m^{-3}

Quantity	Symbol	Symbol for unit
biomass duration: $\int B\,\mathrm{d}t$	Z	kg s
proportion of mass or energy transferred from one particular organism to another	$k, \ldots {}_{i\to j}k$	dimensionless
proportion of mass or energy not transferred from one particular organism to another	$1-k,$ $1-{}_{i\to j}k$	dimensionless
proportion of available food consumed	p	dimensionless
proportion of available food not consumed	$1-p$	dimensionless
leaf area index: S_l/A	L	dimensionless
spatial leaf area 'density': S_l/V	${}_V L$	m^{-1}
plant concentration: B/V	${}_V B, b$	$kg\ m^{-3}$
leaf area ratio: $S_l/m_{\mathrm{D,p}}$ (subscript p for plant)	L_p	$m^2\ kg^{-1}$
specific leaf area: $S_l/m_{\mathrm{D,l}}$	L_l	$m^2\ kg^{-1}$
fraction of mass of an individual organism in a component part or in an organ at a particular time: e.g. $m_{\mathrm{D,l}}/m_{\mathrm{D,p}}$ (leaf mass ratio) (subscripts as for mass above)	w	dimensionless
leaf area duration: $\int S_l\,\mathrm{d}t$	D	$m^2\ s$
fraction of energy of an individual organism in a component part or in an organ at a particular time (subscripts as for energy above)	q	dimensionless
height from ground to top of canopy	h	m
depth of euphotic zone	z_e	m
height above surface of ground or depth below surface of water	z	m
relative height	z/h	dimensionless
amount of change (increase or decrease) taking place in the interval of time $t-t_0 = \Delta t$ in individual mass: $m(t)-m(t_0)$	Δm	kg

Appendix 6

Quantity	*Symbol*	*Symbol for unit*
in individual energy:		
$Q(t) - Q(t_0)$	ΔQ	J
in biomass: $B(t) - B(t_0)$	ΔB	kg
in energy of biomass:		
$\Lambda(t) - \Lambda(t_0)$	$\Delta\Lambda$	J
in numbers: $N(t) - N(t_0)$	ΔN	—

Rates of change with time

instantaneous rates of change
(increase or decrease) (growth rate),

in individual mass: dm/dt	m'	kg s^{-1}
in individual energy: dQ/dt	Q'	J s^{-1}, W
in biomass: dB/dt	B'	kg s^{-1}
in energy of biomass: $d\Lambda/dt$	Λ'	J s^{-1}, W
in numbers: dN/dt	N'	s^{-1}

average rates of change (increase or
decrease) in the interval of time
$t - t_0 = \Delta t$ (mean growth rate),

$$\frac{\Delta x}{\Delta t} = \frac{1}{t - t_0}\int_{t_0}^{t}\frac{dx}{dt}$$

in individual mass: $\Delta m/\Delta t$	g_m	kg s^{-1}
in individual energy: $\Delta Q/\Delta t$	g_Q	J s^{-1}
in biomass: $\Delta B/\Delta t$	g_B	kg s^{-1}
in energy of biomass: $\Delta\Lambda/\Delta t$	g_Λ	J s^{-1}
in numbers: $\Delta N/\Delta t$	g_N	s^{-1}

average rates of change (increase or
decrease) in the interval of time
$t - t_0 = \Delta t$ per unit area
(population or crop growth rate),

in biomass: g_B/A	$_A g_B$	kg m^{-2} s^{-1}
in energy of biomass: g_Λ/A	$_A g_\Lambda$	J m^{-2} s^{-1}, W m^{-2}
in numbers: g_N/A	$_A g_N$	m^{-2} s^{-1}

average rates of change (increase or
decrease) in the interval of time
$t - t_0 = \Delta t$ per unit volume
(population or crop growth rate),

in biomass: g_B/V	$_V g_B$	kg m^{-3} s^{-1}

Quantity	*Symbol*	*Symbol for unit*
in energy of biomass: g_A/V	rg_A	$\text{J m}^{-3} \text{ s}^{-1}$, W m^{-3}
in numbers: g_N/V	rg_N	$\text{m}^{-3} \text{ s}^{-1}$
instantaneous rates of change (increase or decrease) per unit area of *leaf* surface ($+$ stem surface area if photosynthetic) (net assimilation rate; unit leaf rate)		
in individual mass: $\dfrac{1}{S_l}\dfrac{dm}{dt}$	E_m'	$\text{kg m}^{-2} \text{ s}^{-1}$
in individual energy: $\dfrac{1}{S_l}\dfrac{dQ}{dt}$	E_Q'	$\text{J m}^{-2} \text{ s}^{-1}$
in biomass: $\dfrac{1}{S_l}\dfrac{dB}{dt}$	E_B'	$\text{kg m}^{-2} \text{ s}^{-1}$
in energy of biomass: $\dfrac{1}{S_l}\dfrac{dA}{dt}$	E_A'	$\text{J m}^{-2} \text{ s}^{-1}$
average rates of change (increase or decrease) per unit area of *leaf* surface ($+$ stem surface if photosynthetic); in the interval of time $t-t_0 = \Delta t$ (mean net assimilation rate; mean unit leaf rate),		
in individual mass: $\dfrac{1}{S_l}\dfrac{\Delta m}{\Delta t}$	E_m	$\text{kg m}^{-2} \text{ s}^{-1}$
in individual energy: $\dfrac{1}{S_l}\dfrac{\Delta Q}{\Delta t}$	E_Q	$\text{J m}^{-2} \text{ s}^{-1}$
in biomass: $\dfrac{1}{S_l}\dfrac{\Delta B}{\Delta t}$	E_B	$\text{kg m}^{-2} \text{ s}^{-1}$
in energy of biomass: $\dfrac{1}{S_l}\dfrac{\Delta A}{\Delta t}$	E_A	$\text{J m}^{-2} \text{ s}^{-1}$
instantaneous relative (specific) rates of change (increase or decrease) (relative growth rate),		
in individual mass: $\dfrac{1}{m}\dfrac{dm}{dt}$	G_m'	s^{-1}
in individual energy: $\dfrac{1}{Q}\dfrac{dQ}{dt}$	G_Q'	s^{-1}
in biomass: $\dfrac{1}{B}\dfrac{dB}{dt}$	G_B'	s^{-1}

Quantity	*Symbol*	*Symbol for unit*
in energy of biomass: $\dfrac{1}{\Lambda}\dfrac{d\Lambda}{dt}$	G'_Λ	s^{-1}
in numbers: $\dfrac{1}{N}\dfrac{dN}{dt}$	G'_N	s^{-1}
average relative (specific) rates of change (increase or decrease) in the interval of time $t - t_0 = \Delta t$ (mean relative growth rate),		
in individual mass: $\dfrac{1}{m}\dfrac{\Delta m}{\Delta t}$	G_m	s^{-1}
in individual energy: $\dfrac{1}{Q}\dfrac{\Delta Q}{\Delta t}$	G_Q	s^{-1}
in biomass: $\dfrac{1}{B}\dfrac{\Delta B}{\Delta t}$	G_B	s^{-1}
in energy of biomass: $\dfrac{1}{\Lambda}\dfrac{\Delta \Lambda}{\Delta t}$	G_Λ	s^{-1}
in numbers: $\dfrac{1}{N}\dfrac{\Delta N}{\Delta t}$	G_N	s^{-1}
average rates of consumption (intake as food by an individual animal or a population) in the interval of time $t - t_0 = \Delta t$,		
of mass by an individual	C_m	$kg\ s^{-1}$
of energy by an individual	C_Q	$J\ s^{-1}$, W
of biomass	C_B	$kg\ s^{-1}$
of energy of biomass	C_Λ	$J\ s^{-1}$, W
of numbers of another species (j)	C_{N_j}	s^{-1}
(subscripts A per unit ground/ water area: m^{-2}, S per unit surface area of organism: m^{-2}, V per unit volume of habitat or organism: m^{-3}, result in a change in the units, e.g. $_VC_B$ has units of $kg\ m^{-3}\ s^{-1}$)		
average photo-assimilation rate (see page 236) (subscripts as appropriate)	A_p	$mol\ m^{-3}\ s^{-1}$ $(kg\ m^{-2}\ s^{-1})$

Quantities, units and symbols

Quantity	Symbol	Symbol for unit
average rates of loss of the whole or part of a plant or animal (including release of gonadic material and transfer to decomposition food chains) in the interval of time $t - t_0 = \Delta t$,		
* in general	L	kg s^{-1} J s^{-1}, W
average rates of loss of whole or part of a plant or animal to one or more specified predator or herbivore in the interval of time $t - t_0 = \Delta t$ (yield)		
* in general	Y	kg s^{-1} J s^{-1}, W
average rates of loss by faecal excretion (including other egesta) in the interval of time $t - t_0 = \Delta t$		
* in general	F	kg s^{-1} J s^{-1}, W
average rates of loss by urine excretion in the interval of time $t - t_0 = \Delta t$		
* in general	U	kg s^{-1} J s^{-1}, W
average rates of loss by respiration by a plant or animal in the interval of time $t - t_0 = \Delta t$		
* in general	R	kg s^{-1} J s^{-1}, W

(subscripts p pulmonary, c cutaneous, D in darkness, L in light, P photorespiration rate in photosynthetic tissues in light, K respiration rate in non-photosynthetic tissue in light or darkness)

* In each case use subscripts as for consumption (page 242) and also use subscript m, Q, B, Λ to denote mass, energy, biomass or energy of biomass, respectively, as appropriate and necessary. It may not be necessary to repeat subscripts throughout a given text.

Appendix 6

Quantity	Symbol	Symbol for unit

respiratory quotient: $\dot{V}_{CO_2}/\dot{V}_{O_2}$ — no symbol — dimensionless

photosynthetic quotient: $\dot{V}_{O_2}/\dot{V}_{CO_2}$ — no symbol — dimensionless

rate of production (net primary production) by a species or group of species* — P — kg s^{-1}; J s^{-1}, W

   in biomass: $g_B + L_B + Y_B$ — P_B — kg s^{-1}

   in energy of biomass: $g_A + L_A + Y_A$ — P_A — J s^{-1}, W

mean growth rate of an animal

   in mass: $C_m - (F_m + U_m + R_m + L_m)$ — g_m — kg s^{-1}

   in energy: $C_Q - (F_Q + U_Q + R_Q + L_Q)$ — g_Q — J s^{-1}, W

mean rate of assimilation by an animal

   in mass: $C_m - F_m$ — A_m — kg s^{-1}

   in energy: $C_Q - F_Q$ — A_Q — J s^{-1}, W

mean rate of production by an animal

   in mass: $g_m + L_m + U_m$ — P_m — kg s^{-1}

   in energy: $g_Q + L_Q + U_Q$ — P_Q — J s^{-1}, W

efficiency of energy conversion, quantum yield (flux of CO_2 fixed/flux of quanta (photons))

   absorbed: $F_{CO_2}\bigg/\int\int_{\lambda_1}^{\lambda_2}\alpha(\lambda)I_q(\lambda)\,.d\lambda$ — Φ_a — dimensionless (mol Einstein^{-1})

   incident: $F_{CO_2}/I_q(\lambda_1 - \lambda_2)$ — Φ_1 — dimensionless (mol Einstein^{-1})

(mean value for the interval Δt obtained by integrating both sides between t_0 and t)

(subscripts l leaf, p plant, s stem, C canopy, etc.)

*in general: $P = g + L + Y$

rate of production	=	rate of change in biomass or energy	+	rate of non-predatory loss	+	rate of loss to predator or herbivore

Quantity	Symbol	Symbol for unit
chemical energy fixed/incident radiant energy		
$\Delta Q \Big/ \int_{t_0}^{t} I(\lambda_1 - \lambda_2)\,\mathrm{d}t$	$\epsilon_{\mathrm{i},Q}$	dimensionless
$\Delta \Lambda \Big/ \int_{t_0}^{t} I(\lambda_1 - \lambda_2)\,\mathrm{d}t$	$\epsilon_{\mathrm{i},\Lambda}$	dimensionless
chemical energy fixed/absorbed radiant energy		
$\Delta Q \Big/ \int_{\lambda_1}^{\lambda_2} \int_{t_0}^{t} \alpha(\lambda)\, I(\lambda)\,\mathrm{d}\lambda\,\mathrm{d}t$	$\epsilon_{\mathrm{a},Q}$	dimensionless
$\Delta \Lambda \Big/ \int_{\lambda_1}^{\lambda_2} \int_{t_0}^{t} \alpha(\lambda)\, I(\lambda)\,\mathrm{d}\lambda\,\mathrm{d}t$	$\epsilon_{\mathrm{a},\Lambda}$	dimensionless
chemical energy fixed/energy of food consumed		
$\Delta Q \Big/ \int_{t_0}^{t} C_Q\,\mathrm{d}t$	$\epsilon_{\mathrm{c},Q}$	dimensionless
$\Delta \Lambda \Big/ \int_{t_0}^{t} C_\Lambda\,\mathrm{d}t$	$\epsilon_{\mathrm{c},\Lambda}$	dimensionless
rates of change in population numbers in the interval of time $t - t_0 = \Delta t$ expressed as a proportion of the population at time t_0		
birth rate	b	s^{-1}
immigration rate	i	s^{-1}
death rate	d	s^{-1}
capture rate	f	s^{-1}
emigration rate	e	s^{-1}
survival rate: N_t/N_{t_0}	s	s^{-1}
total recruitment rate: $b+i$	r_e	s^{-1}
total mortality rate: $1-s$, $d+f+e$	m	s^{-1}
partial pressure of nitrogen in air	$P_{\mathrm{N_2}}$	Pa

8b. Nitrogen fixation

Quantity	Symbol	Symbol for unit
rate of nitrogen fixation (usually measured as rate of C_2H_2 reduction)	J	mol s^{-1}
in extracts or in whole cells per unit mass of protein	J_p	mol kg^{-1} s^{-1}
in unit volume of fresh water	$_VJ$	mol m^{-3} s^{-1}
in fresh water per mole nitrogen present	J_m	s^{-1}
per unit area of land	$_AJ$	mol m^{-2} s^{-1}
per unit mass of soil	J_g	mol kg^{-1} s^{-1}
nitrogenase protein–complex I	component I	—
nitrogenase protein–complex II	component II	—

9. BIOLOGICAL MATERIAL

Major ranks (phyla, classes, orders, families, etc.) should be written in Roman with an initial capital.

The name of an organism should be specified by the appropriate italicized binomial or trinomial given in full, with attributions, on the first occasion of its use, e.g.

Picea sitchensis (Bong.) Carr.

On subsequent occasions, the generic name may be abbreviated to a single, italicized, capital, and the attributions dropped, e.g.

P. sitchensis

In the case of a trinomial (animals only) in which the specific and subspecific names are identical, the specific name may be abbreviated to the first letter, printed in lower case, italic type, e.g.

Lagopus s. scoticus

Plant subspecies, cultivars and provenances should be indicated as follows:

Veronica serpyllifolia L. subsp. *humifusa* (Dickson) Syme
Pisum sativum L. cv. 'Marrow fat'
Picea sitchensis (Bong.) Carr. provenance Queen Charlotte Islands

The abbreviation of species is sp. (plural spp.).
The abbreviation of subspecies is subsp. (plural subspp.).

The following biochemically based classification may also be used to indicate the photosynthetic pathways:

Plant in which the dominant primary carboxylation is mediated by:	*Designation*
ribulose diphosphate carboxylase	C_3-plant
phosphoenolpyruvate carboxylase in the daytime	C_4-plant
phosphoenolpyruvate carboxylase at night	CAM-plant

10. SOME USEFUL MATHEMATICAL SYMBOLS AND OPERATORS

When letters of the alphabet are used to form mathematical operators (*Examples*: d, Δ, ln, exp) or as mathematical constants (*Examples*: e, π) they should be printed in roman (upright) type so as to distinguish them from the symbols for physical quantities which should be printed in italic (sloping) type.

equal to	$=$	smaller than	$<$
not equal to	$\neq$	larger than	$>$
identically equal to	$\equiv$	smaller than or equal to	$\leqslant$
corresponds to	$\triangleq$	larger than or equal to	$\geqslant$
approximately equal to	$\approx$	much smaller than	$\ll$
approaches	$\rightarrow$	much larger than	$\gg$
asymptotically equal to	$\simeq, \sim$	plus	$+$
proportional to	$\propto$	minus	$-$
infinity	∞	plus or minus	$\pm$

a multiplied by b $\qquad ab, a \cdot b, a \times b$

a divided by b $\qquad a/b, \dfrac{a}{b}, ab^{-1}$

magnitude of a $\qquad |a|$

a raised to power n $\qquad a^n$

square root of a $\qquad a^{\frac{1}{2}}, \sqrt{a}$

nth root of a $\qquad a^{1/n}, a^{\frac{1}{n}}, n\sqrt{a}$

mean value of a $\qquad \langle a \rangle, \bar{a}$

factorial p* $\qquad p!$

binomial coefficient† $\qquad \binom{n}{p}$

sum $\qquad \Sigma$

product $\qquad \Pi$

function of x $\qquad f(x), \mathrm{f}(x)$

limit to which $f(x)$ tends as x approaches a $\qquad \lim_{x \to a} f(x), \lim_{x \to a} f(x)$

finite increment of x $\qquad \Delta x$

variation of x $\qquad \delta x$

differential coefficient of $f(x)$ with respect to x $\qquad \dfrac{\mathrm{d}f}{\mathrm{d}x}, \mathrm{d}f/\mathrm{d}x, f'(x)$

---

* $p! = 1 \times 2 \times 3 \times \ldots \times (p-1) \times p$ where p is a positive integer.

† $\binom{n}{p} = n!/(n-p)!p!$ where n and p are positive integers and $n \geqslant p$ and where $0! = 1$.

| differential coefficient of order n of $f(x)$ | $\dfrac{\mathrm{d}^n f}{\mathrm{d}x^n}$, $\mathrm{d}^n f/\mathrm{d}x^n$, $f^{(n)}(x)$ |

partial differential coefficient of $f(x, y, ...)$ with

respect to x when $y, ...$ are held constant $\dfrac{\partial f(x, y, ...)}{\partial x}$, $\left(\dfrac{\partial f}{\partial x}\right)_y$, $(\partial f/\partial x)_y$

the total differential of f — $\mathrm{d}f$

indefinite integral of $f(x)$ with respect to x — $\int f(x)\,\mathrm{d}x$

definite integral of $f(x)$ from $x = a$ to $x = b$ — $\int_a^b f(x)\,\mathrm{d}x$

integral of $f(x)$ with respect to x round a closed contour — $\oint f(x)\,\mathrm{d}x$

exponential of x	$\exp x$, e^x
base of natural logarithms	e
logarithm to the base a of x	$\log_a x$
natural logarithm of x	$\ln x$, $\log_e x$
common logarithm of x	$\log_{10} x$, $\log x$, $\lg x$
ratio of circumference to diameter of a circle	π
sine of x	$\sin x$
cosine of x	$\cos x$
tangent of x	$\tan x$
cotangent of x	$\cot x$
secant of x	$\sec x$
cosecant of x	$\operatorname{cosec} x$
gradient of ϕ	$\operatorname{grad} \phi$, $\nabla\phi$
divergence of A	$\nabla.\mathbf{A}$, $\operatorname{div}\mathbf{A}$

Statistical symbols

Greek letters are used as parameters and italicized lower case letter for estimated values.

standard deviation	σ
estimated standard deviation	s, s.d.
variance	σ^2
estimated variance or mean square	s^2
degrees of freedom	d.f.
standard error	s.e.
simple correlation coefficient	r
multiple correlation coefficient	R
partial correlation coefficient	$r_{xy.z}$
coefficient of variation s/x	c.v.
coefficient of determination	r^2, R^2
mean of x	$\bar{x}$

Appendix 6

11. SOME USEFUL PHYSICAL CONSTANTS AND QUANTITIES

Constant quantity	Symbol	Unit
gas constant	$R = Lk$	8.31434 J K^{-1} mol^{-1}
	$RT_{273.15K}$	2.27106×10^3 J mol^{-1}
molar volume of ideal gas at 101 325 Pa (1 atm) and 273.15 K	V_m	2.24136×10^{-2} m^3 mol^{-1}
Planck constant	h	6.626196×10^{-34} J s
Boltzmann constant	$k = R/L$	1.38062×10^{-23} J K^{-1}
Avogadro constant	N_A, L	6.02217×10^{23} mol^{-1}
ice point	T_{ice}	273.150 K
Planck constant	h	6.6256×10^{-34} J s
velocity of light	c_0	2.997925×10^8 m s^{-1}
Stefan–Boltzmann constant	σ	5.6696×10^{-8} W m^{-2} K^{-4}

Mathematical constants

ratio of circumference to diameter of a circle	π	3.14159
	π^{-1}	0.31830
	π^2	9.8696
base of natural logarithms	e	2.76828
$\log_{10}$ e	M	0.43429
	M^{-1}	2.30258
acceleration due to gravity (at N 51° 28.1′ Greenwich)	g	9.819 m s^{-2}
solar constant		≈ 1360 W m^2

Values of physical quantities at 20 °C and standard pressure 101 325 Pa

Quantity	Symbol	Value
Air		
molar mass	M_a	≈ 28.97 kg kmol^{-1}
specific heat at constant pressure	c_p	1.01×10^3 J kg^{-1} K^{-1}
thermal diffusivity	D_h	0.208×10^{-4} m^2 s^{-1}
thermal conductivity	k_h	0.0257 J m^{-1} s^{-1} K^{-1}
density	ρ_a	1.205 kg m^{-3}

Quantities, units and symbols

dynamic viscosity μ_a 0.181×10^{-4} Pa s
kinematic viscosity ν_a 0.150×10^{-4} m² s⁻¹

Water

latent heat of vaporization L, λ 2.454×10^6 J kg⁻¹
specific heat at constant pressure c 4.182×10^3 J kg⁻¹ K⁻¹
dynamic viscosity μ 10.2×10^{-4} Pa s
thermal conductivity k_h 0.59 J m⁻¹ s⁻¹ K⁻¹
surface tension r, σ 72.75×10^{-3} kg s⁻²
diffusivity of water vapour in air D_{H_2O}, D_v 0.255×10^{-4} m² s⁻¹

Carbon dioxide

density ρ_{CO_2} 1.842 kg m⁻³
solubility 36.5 mol m⁻³
diffusivity in air D_{CO_2} 0.15×10^{-4} m² s⁻¹
diffusivity in water (25 °C) 0.17×10^{-8} m² s⁻¹

Oxygen

density ρ_{O_2} 1.331 kg m⁻³
solubility 1.36 mol m⁻³
diffusivity in air D_{O_2} 0.191×10^{-4} m² s⁻¹
diffusivity in water (25 °C) 0.292×10^{-8} m² s⁻¹
maximum content in water saturated with atmospheric air 6.36 dm³ m⁻³

251

12. CONVERSION TABLES

acre	4046.86 m²
are, a	100 m²
atmosphere, standard; atm	101325 Pa
bar	10⁵ Pa
British thermal unit, Btu	1.05506 kJ
British thermal unit foot/square foot hour	
degree F, (Btu ft)/(ft² h F)	1.73073 W m⁻¹ K⁻¹
British thermal unit/hour, Btu/h	0.293071 W
bushel, bu	0.0363687 m³
calorie, thermochemical	4.184 J
calorie centimetre/square centimetre second	
degree C, (cal cm)/(cm² s °C)	418.4 W m⁻¹ K⁻¹
Celsius degree, °C	$(T/°C + 273.15)$ K
cubic foot, ft³	0.0283168 m³
cubic inch, in³	16.3871 cm³
cubic yard, yd³	0.764555 m³
curie, Ci	37×10^9 s⁻¹
cusec (cubic foot per second)	0.0283168 m³ s⁻¹
degree Celsius, °C	$(T/°C + 273.15)$ K
degree Fahrenheit, °F	$5/9\,(T/°F + 459.67)$ K
dram (avoirdupois), dr	1.77185 g
dyne, dyn	10⁻⁵ N
electron volt, eV	1.60210×10^{-19} J
erg	10⁻⁷ J
fathom	1.8288 m
fluid ounce, fl oz	28.4131 cm³
foot, ft	0.3048 m
foot-candle, lm/ft²	10.7639 lx
foot/minute, ft/min	0.00508 m s⁻¹
foot of water (conventional pressure unit)	2989.07 Pa
gallon, gal	4.54609 dm³
gallon (US), USgal	3.78541 dm³
gill	0.142065 dm³
grain, gr	64.7989 mg

hectare, ha	10^4 m^2
horsepower, hp	745.700 W
horsepower hour, hp h	2.68452 MJ
hour, h	3600 s
hundredweight, cwt	50.8023 kg
inch, in	25.4 mm
inch of mercury	3386.39 Pa
inch of water	249.089 Pa
international nautical mile, n mile	1.852 km
kilocalorie/hour, kcal/h	1.163 W
kilocalorie/square metre hour, kcal/(m^2 h)	1.163 W m^{-2}
kilocalorie/square metre hour degree C, kcal/(m^2 h °C)	1.163 W m^{-2} K^{-1}
knot, international: kn (1 n mile/h)	1.852 km h^{-1}
knot UK	1.853184 km h^{-1}
litre atmosphere, l atm	101.328 J
lumen/square foot, lm/ft^2	10.7639 lx
mile	1.60934 km
mile, international nautical; n mile	1.852 km
mile, nautical UK	1.853184 km
millibar, mbar	100 Pa
millimetre of mercury, mmHg	133.322 Pa
millimetre of water	9.80665 Pa
minim	59.1939 mm^3
minute (time), min	60 s
molar, mol/l, m	1 mol dm^{-3}
ounce, oz	0.0283495 kg
ounce, fluid; fl oz	28.4131 cm^3
perch	5.0292 m
pint, pt	0.568261 dm^3
poise, P	0.1 Pa s
poiseuille, Pl	1 Pa s
pole	5.0292 m
pound, lb	0.45359237 kg
pound-force, lbf	4.44822 N
pound-force/square inch, lbf/in^2 (or p.s.i.)	6894.76 Pa

quart, qt	1.136 52 dm³
rad (100 erg/g)	0.01 J kg⁻¹
rod	5.029 2 m
square foot, ft²	0.092 903 0 m²
square inch, in²	645.16 mm²
square mile, sq mile	2.589 99 km²
square yard, yd²	0.836 127 m²
stokes, St	10^{-4} m² s⁻¹
thou	25.4 μm
ton	1 016.05 kg
tonne (metric ton), t	1 Mg
ton of refrigeration	3 516.85 W
torr	133.322 Pa
yard, yd	0.914 4 m

13. REFERENCES

1 *Manual of Symbols and Terminology for Physiochemical Quantities and Units.* Ed. M. L. MCGLASHAN for the International Union of Pure and Applied Chemistry, Butterworths, London, 1971.

2 *Metric Units, Conversion Factors and Nomenclature in Nutritional and Food Sciences.* The Royal Society, London, 1972.

3 Proposed Standard System of Symbols for Thermal Physiology. *Journal of Applied Physiology*, **27**, 439–46, 1969.

4 *Quantities, Units and Symbols.* The Symbols Committee of The Royal Society, 1971. (New edition in preparation.)

5 *The International System of Units (SI).* National Physical Laboratory UK, 1973.

6 *SI and Metrication Conversion Tables.* G. SOCRATES and L. J. SAPPER, Newnes–Butterworths, London, 1969.

7 *Symbols, Units and Conversion Factors in Studies of Freshwater Productivity.* G. C. WINBERG and collaborators, IBP, London, 1971.

8 *Tentative Recommendations of Terminology, Symbols and Units in Plant Physiology.* Ed. H. BURSTRÖM for the International Association for Plant Physiology, Lund, 1972.

9 *The International System of Units (SI).* British Standards Institute, 1970.

10 *SI Units and Recommendations for the use of their Multiples and of Certain Other Units* ISO 1000 1973 (also available in French).

11 *General Principles Concerning Quantities, Units and Symbols* ISO/R 31/0 1973 (also available in French).

List of abbreviations used in text

ACMRR	Advisory Committee on Marine Resources Research
CODATA	Committee on Data for Science and Technology
COWAR	Committee on Water Research
CT	Conservation Terrestrial
FAO	UN Food and Agriculture Organisation
FASEB	Federation of American Societies for Experimental Biology
GIAM	Global Implications of Applied Microbiology
HA	Human Adaptability
IAA	International Association of Aerobiology
IABO	International Association of Biological Oceanography
IAHB	International Association of Human Biologists
IAPP	International Association of Plant Physiologists
IBP	International Biological Programme
ICES	International Council for the Exploration of the Sea
ICNAF	International Council for North Atlantic Fisheries
ICRO	International Cell Research Organization
ICSU	International Council for Scientific Unions
IGU	International Geographical Union
IGY	International Geophysical Year
IHD	International Hydrological Decade
IHP	International Hydrological Programme
INTECOL	International Association for Ecology
IOC	Intergovernmental Oceanographic Commission
ISSS	International Society of Soil Scientists
IUAES	International Union of Anthropological and Ethnological Sciences
IUB	International Union of Biochemistry
IUBS	International Union of Biological Sciences
IUCN	International Union for the Conservation of Nature and Natural Resources
IUFRO	International Union of Forest Research Organizations
IUGG	International Union of Geodesy and Geophysics
IUNS	International Union of Nutritional Sciences
IUPAB	International Union for Pure and Applied Biophysics
IUPS	International Union of Physiological Sciences

IUSSI	International Union for Study of Social Insects
LEPOR	Long-term and Expanded Programme of Oceanic Exploration and Research
MAB	Man and the Biosphere Programme
PAWE	Program for Analysis of World Ecosystems
PF	Productivity Freshwater
PM	Productivity Marine
PP	Production Processes
PSA	Pacific Science Association
PT	Productivity Terrestrial
SCAR	Scientific Committee on Antarctic Research
SCIBP	Special Committee for the International Biological Programme
SCOPE	Scientific Committee on Problems of the Environment
SCOR	Scientific Committee on Oceanic Research
SIL	International Association of Limnology
UM	Use and Management
UNDP	United Nations Development Programme
UNEP	United Nations Environment Programme
UNESCO	United Nations Educational, Scientific and Cultural Organisation
UNO	United Nations Organisation
WHO	World Health Organisation
WMO	World Meteorological Organisation

References

Ackermann, W. C., White, G. F. & Worthington, E. B. (eds.) (1973). *Man-Made Lakes: their Problems and Environmental Effects.* American Geophysical Union, Washington.

Allen, O. N., Hamatová, E. & Skinner, F. A. (1973). *IBP World Catalogue of Rhizobium Collections.* IBP Central Office, London.

Baker, P. T. & Weiner, J. S. (1966). *The Biology of Human Adaptability.* Clarendon Press, Oxford.

Clarke, G. L. (1946). Dynamics of production in a marine area. *Ecological Monographs,* **16,** 321–35.

Cody, M. L. (1974). Optimization in ecology. *Science,* **183,** 1156–64.

Dasmann, R. F., Milton, J. P. & Freeman, P. H. (1973). *Ecological Principles for Economic Development.* John Wiley, Chichester, for IUCN and the Conservation Foundation.

Dobzhansky, T. (1962). *Mankind Evolving.* Yale University Press, New Haven, Conn.

Emlen, J. M. (1973). *Ecology: An evolutionary approach.* Addison-Wesley, Reading, Mass.

Frankel, O. H. (1972). Australia and the International Biological Programme. *Search,* **3** (4), 105–8.

Frankel, O. H. & Bennett, E. (eds.) (1970). *Genetic Resources in Plants – their Exploration and Conservation,* IBP handbook No. 11. Blackwell, Oxford.

Harrison, G. A., Weiner, J. S., Tanner, J. M. & Barnicot, N. A (1964). *Human Biology: An Introduction to Human Evolution, Variation and Growth.* Clarendon Press, Oxford.

Harrison, G. A. & Boyce, A. J. (eds.) (1972). *The Structure of Human Populations.* Clarendon Press, Oxford.

IUBS (International Union of Biological Sciences) (1961). *Proceedings: IUBS General Assembly.*

Ivlev, V. S. (1945). (The biological productivity of water). *Uspekhi Sovremennoi Biologii,* **19,** 98–120 (In Russian).

Jeffers, J. N. R. (ed.) (1972). *Mathematical Models in Ecology,* 12th *Symposium of the British Ecological Society.* Blackwell, Oxford.

Lindeman, R. L. (1942). The trophic–dynamic aspect of ecology. *Ecology,* **23,** 399–418.

259

References

Lowe-McConnell, R. H. (ed.) (1966). *Man-Made Lakes*, Institute of Biology Symposia No. 15. Academic Press, London.

MacArthur, R. H. (1972). *Geographical Ecology*. Harper & Row, New York.

Macfadyen, A. (1948). The meaning of productivity in biological systems. *Journal of Animal Ecology*, **17**, 75–80.

Macfadyen, A. (1957). *Animal Ecology: Aims and Methods*. Pitman, London.

Margalef, R. (1968). *Perspectives in ecological theory*. University of Chicago Press, Chicago.

May, R. M. (1973). *Stability and complexity in model ecosystems*. Princeton University Press, Princeton, N.J.

Maynard Smith, J. (1974). *Models in Ecology*. Cambridge University Press, London.

Montalenti, G. (1961). The international biological programme. *ICSU Review*, **3** (2), 72–7.

Montalenti, G. (1966). The International Biological Programme. *Penguin Science Survey*, B, 181–8.

Munn, R. E. (1973). *Global environmental monitoring system (GEMS). Action plan for phase I*. SCOPE Report 3, Toronto.

Obeng, L. E. (ed.) (1969). *Man-Made Lakes*, The Accra Symposium. Ghana University Press, Accra.

Odum, E. P. (1953). *Fundamentals of Ecology*. Saunders, Philadelphia.

ᴊdum, H. T. (1971). *Environment, Power and Society*. Wiley Interscience, New York.

Poore, M. E. D. & Robertson, V. C. (1964). *An approach to the rapid description and mapping of biological habitats*. IBP Central Office, London.

Ricklefs, R. E. (1973). *Ecology*. Nelson, London.

SCOPE (Scientific Committee on Problems of the Environment) (1971). *Global Environmental Monitoring*, Report No. 1. International Council of Scientific Unions, Paris.

SCOPE (Scientific Committee on Problems of the Environment) (1972). *Man-Made Lakes as Modified Ecosystems*, Report No. 2. International Council of Scientific Unions, Paris.

Simon, N. (1966). *Mammalia. IUCN Red Data Book Vol. I*. IUCN, Lausanne.

Sorre, M. (1943). *Les fondements biologiques de la géographie humaine*. Armand Collin, Paris.

Thienemann, A. (1931). Der Produktionsbegriff in der Biologie. *Archiv für Hydrobiologie*, **22,** 616–22.

United Nations (1971). *Report of the First Session of the Intergovernmental Working Group on Monitoring and Surveillance.* Geneva.

UNESCO (United Nations Educational, Scientific and Cultural Organization) (1972*a*). *Expert panel on the role of systems analysis and modelling approaches in the Programme on Man and the Biosphere (MAB). Final Report.* Paris.

UNESCO (United Nations Educational, Scientific and Cultural Organization) (1972*b*). *Expert panel on project 12: interactions between environmental transformations and genetic and demographic changes in the Programme on Man and the Biosphere (MAB). Final Report.* Paris.

Van Dyne, G. M. (ed.) (1969). *The ecosystem concept in natural resource management.* Academic Press, New York.

Vincent, J. (1966). *Aves. IUCN Red Data Book Vol. II.* IUCN, Lausanne.

Watt, K. E. F. (1968). *Ecology and resource management: a quantitative approach.* McGraw-Hill, New York.

Index

actinomycetes, as nitrogen fixers, 27
Advisory Committee on Marine Resources Research (ACMRR), 111
aerobiology, 108, 113
agroclimatology, FAO/UNESCO/WMO Working Group on, 26
agrosystems, 132–3
Ainu people, 98, 99
algae: culture of, 49; in fresh waters, 95; excretion of carbon compounds by, 37
'animal weeds', 72
anthropometry, 97
ants, 71
aquaculture, 93
aquatic macrophytes, production and ecology of, 89
Arctic: collaboration of countries on, 59; people of, 98, 99, 109
Argentina, 95, 157, 190
Arid Zones and Humid Tropics programme (UNESCO), 26
Aridland Biome, 68, 107, 112
Atlantic (North), fisheries of, 93
Australia, 101, 157, 190
Austria, 157, 190
autecology, 63

bacteria: as food for other organisms, 92; in fresh waters, 37; nitrogen-fixing (nitrifying), 27–8, 41, 78–9; in sea water, 41
Baer, Jean, 4, 5, 6, 12, 56
Baker, Paul, 99
Belgium, 11, 101, 157, 190; joint projects of, with India and Zaïre, 100
benthos, methods for studying, 94
Bhutan, 101
biomes, 65, 66, 136; cartoon of research on, 70; stations for research on, 85
Bioscience, IBP topics in, 123
Biosphere, The, bulletin of IBP, 118, 181
Biosphere Conference (UNESCO, 1968), xvii, 118
birds: granivorous, 71–2, 108, 112, 134; inland waters of importance to, 92
Blackman, G. L., 76
blood sampling and collection, 97
blue-green algae, as nitrogen fixers, 27, 41, 78, 79
Bourlière, F., 8, 56, 131
Brazil, 100, 157, 190
Britain, *see* United Kingdom

Brown, Sir Lindor, 4, 14
Bulgaria, 157, 191
Byerly, T. C., 9

Canada, 59, 84, 95, 158; publications in, 105, 120, 191
canals, inter-oceanic, 94
carbon cycle, 24
Central Office of IBP, xix–xx, 61, 63, 121, 126; expenses of, 128; staff of, 152
Chad, 101
Check Sheet Survey of Areas, 32, 80–5, 111
chickpea (*Cicer arietinum*), 47
Chile, 158
circumpolar people, 109
Clark, R. B., 95, 111
climatic zones, 34–5
cold tolerance, in humans, 97, 98
Colombia, 158
Committee for Data in Science and Technology (CODATA), 60
Committee on Water Research of ICSU (COWAR), 110, 112
Commonwealth Agricultural Bureaux, 46
Commonwealth Foundations, 128
conservation, 52–3, 108; of genetic resources, 30, 88–9; *see also* conservation section
Conservation Foundation, 85
Conservation of Terrestrial Communities (CT) section, 7, 12–14; committee of, 154; operations of, 80–6; programme of, 30–3; transfer of check survey and data bank of, 113, 137
conveners of Sectional Committees, 15, 56, 57, 125, 152–6
Cragg, J. B., 65
crop plants: collections of, 46; genetic resources of, 108, 113; wild varieties and primitive cultivars of, 45, 46, 87, 88
Czechoslovakia, 53, 89, 101, 158; PP-P subsection in, 76, 126; publications in, 191

Danube, delta of, 89
data: collection and processing of, 104; storage and retrieval of, 19, 60, 73–4, 82, 83, 84
data bank, of CT section, 113, 187
decomposition processes, 24, 71, 108, 113
demographic surveys, 97
Denmark, 59, 158, 191; *see also* Scandinavia
deserts, human communities in, 98

263

Index

developing countries: and IBP, xviii, 53–4, 59, 157; inoculation of legume seeds in, 79; joint projects of scientifically developed countries and, 59, 100
development projects, environmental considerations in, 85
diatoms, 41
Dubos, René, 66
Dunbar, Max, 93
Dunn, L. C., 4
Duvigneaud, P., 11

East Africa, 59, 158, 191; see also Kenya, Tanzania, Uganda
Ebert, T. A., 10
Eckardt, F. E., 75
ecology, xx, 131; becoming a predictive science, 134; in USA, 9–10
Ecophysiology, UNESCO Commission for, 75
ecosystems, 25–6; approach through, 63–4, 132; conservation of, 108, 113; dynamics of, 68–9, 71; inventory of, 31–2; loss of, 52; 'natural', 15, 22
Ecuador, 158, 192
editors, of international synthesis series, 104, 107, 109
education and training: for biological control of pests, 48; for environmental science, 144; for freshwater biology, 35; for IBP work, 19; in IBP work, 72–3; for taxonomic work with ecological outlook, 146
Egypt, 100, 101, 159
Eichhornia crassipes (water hyacinth), 90
Ellenberg, H., 6, 11, 13, 82
Engelhardt, 4, 5
environment: developmental projects and, 85; monitoring of changes in, 110–11, 135–7; in relation to production, 24–5; United Nations conference on (Stockholm 1972), xviii, 88, 111
Eskimo people, 98, 99

FAO, 79, 100; collaboration of IBP and, 46, 85, 87; and fisheries research, 93; transfer of IBP work to organisations of, 112, 113, 114, 136
farm animals, foods for, 49–50
Farner, Don, 103
Fenn, Wallace, 14, 98
finance of IBP, xviii, 13, 54, 55–6, 125–9; contributions to, by participating countries, 166–8; of HA section, 99; of publications, 106, 117, 118, 127–8; of transfer, 114
finance committee of SCIBP, 151–2
Finland, 59, 159, 192; see also Scandinavia

fisheries, 36, 40, 93
fishponds, 35
flagellates, in sea water, 41
Florkin, M., 4
food chains, 23–4, 50
Ford Foundation, 85, 127, 128
Forest (deciduous) Biome, 68
Fosberg, F. R., 82
France, 8, 59, 95, 192; joint project of Senegal and, 100
Francis of Assisi, St, 1–2
Frankel, Sir Otto, 86
freshwater ecosystems, 108, 112; see also productivity (fresh-water) section

Galapagos Islands, Charles Darwin research station in, 85
General Assemblies of IBP, 115; I (Paris 1964), 15, 16, 17, 81, 125, 141, 179; II (Paris 1966), 141, 179; III (Varna 1968), 110, 179; IV (Rome 1970), 143, 179; V (Seattle, 1972) 144, 179
genetic resources: conservation of, 30, 88–9; of plants, 45–6, 86–9, 108, 113
genetics, of nitrogen fixation in plants and bacteria, 79
genotype, and response to environment, 47, 77–8
Germany, Democratic Republic, 159, 192
Germany, Federal Republic, 159, 192
Ghana, 110, 159, 192
Glover, R., 8
Graham, Edward, 6, 9, 12
Grasslands Biome, 68, 107, 112; expenditure on research on, 129
grazing: food chain in, 23; regulation of, 49–50, 72
Greece, 101, 159

habitats, loss of, 52
Hamatová, E., 74
Hart, Sandy, 99
heat tolerance, of humans, 97, 98
herbivores, large, 49, 73
high-altitude peoples, 109
Holt, Sidney, 7, 111
Hulings, Neil, 94
Human Adaptability (HA) section, 7, 14–16, 109; Committee of, 155–6; operations of, 96–102; programme of, 42–5
human biology, IBP and, 134–5
humus, formation of, 24
Hungary, 160, 192

IBP Directory, 182
IBP Handbooks, 17, 19, 58, 98, 115, 119–20, 128, 182–3

Index

mullets, 108, 113
mussels, 108, 113

National Academy of Sciences, USA; grant to IBP from, 127
national activities in IBP: in finance, 126, 127, 128–9; in preparations, 54–6, 58–60; in publications, 17, 104–5, 122, 190–9; in survey of areas, 84–5
Natural Environment Research Council (UK), 111
Nature Conservancy (UK), 8, 11, 13, 126
Netherlands, 96, 161, 195
New Guinea, 101
New Zealand, 94, 96, 161, 195
Nicholson, E. M., 6, 12, 81
Nigeria, 101, 161, 195
Nile, Sudd region of, 89
nitrifying bacteria, 27–8, 41, 78–9
nitrogen cycle, 24, 75
nitrogen fixation, 108, 112; application of minerals and, 80; in freshwater eco-systems, 37; in non-leguminous plants, 28, 79; PP-N subsection on, 27–8, 74, 75, 78–80
Norway, 59, 96, 162, 195; see also Scandinavia
Nuffield Foundation, 127, 128

operations of IBP (Phase 2), 51, 57, 63–5, 143; see also individual sections
Oren, O. H., 95
organic substances: in fresh waters, 37; in sea water, 39, 41

Pacific Islands, list of conservation opportunities and problems in, 85
Pacific Science Association, 59
Pacific Science Congress, 85
Pakistan, 101
Panama, 162
Pantin, C., 4
parasites, 23
Passer spp., 71
Peru, 100, 162
pests, biological control of, 47–8, 86, 108, 113
Peters, Sir Rudolph, 1, 3, 4
Petrusewicz, K., 132
Philippines, 96, 162, 196
phosphate metabolism, 29
photosynthesis, 39, 108, 112; C_3 and C_4 types of, 78; handbook on, 58, 187; PP-P subsection on, 28–9, 74, 75, 76–8
Photosynthetica (Prague), new international journal, 77, 121
Physiological Anthropometry, IUPS Commission of, 98

phytoplankton, 39, 40
plankton statistical project, 94
Plant Exploration and Introduction, FAO Panel on, 87–8
Poland, 53, 67, 89, 162, 196; joint project of Egypt and, 100
politics, and IBP, 101–2
pollution, 111, 136; of fresh waters, 33; of the sea, 93, 95, 111
population dynamics, 71, 72, 133, 134; human, 42–3
population structure, human, 109
Poupa, Ottokar, 101
predators, 23
preparations for IBP (Phase 1), 51, 60–1; achievements and failings in, 56, 57–60; international organisation in, 56; national activities in, 54–6; pro and anti-IBP camps in, 51–4
primary production, 22–3; estimate of global, 25; in freshwater systems, 92
Production Processes (PP) section, 7; Committee of, 153–4; future research in, 146–7; operations of, 74–80; programme of, 26–30; see also nitrogen fixation, photosynthesis
productivity and human welfare, subject of IBP, 5–6, 7, 18, 52, 131, 132
Productivity Freshwater (PF) section, 7; Committee of, 154–5; handbooks of, 119–20; operations of, 90–3; programme of, 33–7
Productivity Marine (PM) section, 7; Committee of, 155; operations of, 93–6; programme of, 37–42
Productivity Terrestrial (PT) section, 7, 13; Committee of, 152–3; operations of, 65–74; programme of, 21–6
programme of IBP, 17–18; general principles for, 18–21; minimum and maximum, 65, 73; see also under individual sections
Project Aqua, 34, 92, 111, 112, 119
Project Mar, 92
projects of work under IBP, 17, 21, 56; list of, 63
protein: efficient production of, 75; food sources of, 26, 48–9, 108, 113; novel sources of, 75; production of, by plants, 30
public relations: proposed section on training and, 7–8, 13, 54; publication in category of, 122–3
publications of IBP, 17, 20–1, 117–23, 181–2; editorial committee for, 51, 152; financing of, 127–8; handbooks, etc, 17, 19, 58, 119–20, 182–4; income from, 126, 128; international synthesis series, 105–9, 115, 122, 119–201; international volumes,

publications of IBP—*contd.*
184–90; national programmes and reports, 115, 122, 190–9

quantities, units, and symbols, report on, 72, 106–7, 203–55

radio-isotopes, 36, 39
research stations: on different biomes, 85; on inland waters, 91, 93; requests from directors of, 144
reserves, national parks, research areas, 30–1, 52; management of, 33
Revelle, Roger, 10
Rhizobium, 27, 28, 78
Rhode, W., 8, 35
Rhodesia, 162, 196
rice, 47; pests of, 48
Rockefeller Foundation, 46
Romania, 89, 95, 162, 196
Royal Society, London, 4, 7, 8, 53, 126
Rzóska, J., 90

salmon fisheries, 36
Salvinia molesta (water fern), 90
Scandinavia, 8, 53, 59, 75, 99, 105; publications in, 196–7; *see also* Denmark, Finland, Norway, Sweden
Science Journal, IBP topics in, 123
Scientific Committee on Antarctic Research (SCAR), 26, 34, 93, 99
Scientific Committee on Oceanic Research (SCOR), 93, 94, 113
Scientific Committee on Problems of the Environment (SCOPE), 110, 111, 112, 113, 142, 143
scientific co-ordinators in IBP, 57, 61, 125, 138, 152–5
Scientific Director of IBP, 11, 149
scientists, IBP network of, xvii, 16, 60, 67, 80, 106, 147
Sea-Ice Conference (1971), 94
Seals, Conference on (1972), 94
secondary production, 23–4, 30
sections of IBP, 7–8, 21, 63, 66n, 115, 151; Committees of, 56–7, 152–6; finance of activities of, 128; programmes of, 18, 138; *see also individual sections*
Senegal, 100
Smith, Frederick E., 10
soil data, 25
solar radiation, 25, 28–9, 39
Solomon Islands, 101
South Africa, 163, 197
Spain, 163, 197
Special Committee for IBP (SCIBP), 11, 51, 56, 100, 103, 105, 107; and finance, 126; meetings of, 179; officers and

members of, 149–50; and publications, 120, 121–2
species, conservation of, 30, 32, 52, 145
Sri Lanka, 163
statistical methods, 40; in plankton research, 94
Stebbins, G. Ledyard, 4, 5, 7, 100
Steele, J. H., 96
Steinberg, 7, 100
Survival Service Commission of IUCN, 32
Sweden, 8, 59, 118, 163; *see also* Scandinavia
symbols, quantities, and units, report on, 72, 106–7, 203–55
synecology, 63
synthesis and transfer stage of IBP (Phase 3), 51, 57, 143; synthesis, 103–9; transfer, 109–16
systems analysis, 52, 64–5, 67–8, 72, 74, 133

Taiwan, 163
Tanzania, 59, 100, 163, 191; Serengeti Research Station in, 85
termites, 71
Thailand, 163, 197
themes of work under IBP, 21, 56, 66, 73
Thorson, Gunnar, 94
Tokalav Islands, 101
Tonolli, L., 8, 90
toxins: chemical, 37; fungal, 49
Tristan da Cunha, 101
tropical areas: fresh water in, 35; humid, 98; mineral deficiencies in soils of, 49; minimum research programme for, 73; poorly known ecology of, 132
Tundra Biome, 66, 68, 69, 107, 112; expenditure on research in, 129; flow diagram of ecosystem of, 69

Uganda, 59, 164, 191
UNESCO, 73, 93, 100; contracts of IBP with, 126, 127; support to IBP from, 138; transfer of IBP work to organisations of, 112, 113
UNESCO/FAO project of mapping world vegetation, 26, 82
United Kingdom (UK), 95, 101, 164; joint projects of, with Malawi and Tanzania, 100; preparations for IBP in, 7, 8, 53; publications in, 105, 197–8
United Nations Development Programme, 110
United Nations Environment Programme (UNEP), xviii, 89
units, quantities, and symbols, report on, 72, 106–7, 203–55
upwelling in oceans, 95
Uruguay, 164